THE EVOLUTION-CREATION STRUGGLE

The
Evolution–Creation
Struggle

MICHAEL RUSE

HARVARD UNIVERSITY PRESS

Cambridge, Massachusetts
London, England

First Harvard University Press paperback edition, 2006

Library of Congress Cataloging-in-Publication Data

Ruse, Michael.
The evolution-creation struggle / Michael Ruse.
p.   cm.
Includes bibliographical references (p.   ) and index.
ISBN-13: 978-0-674-01687-3 (cloth)
ISBN-10: 0-674-01687-4 (cloth)
ISBN-13: 978-0-674-02255-3 (pbk.)
ISBN-10: 0-674-02255-6 (pbk.)
1. Human evolution—Religious aspects—Christianity.   2. Religion and science.   I. Title.
BT712.R88 2005
231.7'652—dc22   2005040282

For Robert J. Richards

Contents

The sea is calm to-night.
The tide is full, the moon lies fair
Upon the straits; —on the French coast the light
Gleams and is gone; the cliffs of England stand,
Glimmering and vast, out in the tranquil bay.
Come to the window, sweet is the night air!
Only, from the long line of spray
Where the sea meets the moon-blanch'd land,
Listen! you hear the grating roar
Of pebbles which the waves draw back, and fling,
At their return, up the high strand,
Begin, and cease, and then again begin,
With tremulous cadence slow, and bring
The eternal note of sadness in.

Sophocles long ago
Heard it on the Aegean, and it brought
Into his mind the turbid ebb and flow
Of human misery; we
Find also in the sound a thought,
Hearing it by this distant northern sea.

The Sea of Faith
Was once, too, at the full, and round earth's shore
Lay like the folds of a bright girdle furl'd.
But now I only hear
Its melancholy, long, withdrawing roar,
Retreating, to the breath
Of the night-wind, down the vast edges drear
And naked shingles of the world.

Ah, love, let us be true
To one another! for the world, which seems
To lie before us like a land of dreams,
So various, so beautiful, so new,
Hath really neither joy, nor love, nor light,
Nor certitude, nor peace, nor help for pain;
And we are here as on a darkling plain
Swept with confused alarms of struggle and flight,
Where ignorant armies clash by night.

Prologue

$\mathcal{M}$atthew Arnold's great poem "Dover Beach" was written on his honeymoon in 1852, but it was not published until 1867. Between those dates, in 1859, the English naturalist Charles Darwin published *On the Origin of Species.* In that book, Darwin argued that all organisms, including us humans, are the end result of a long, slow process of development—of evolution—from forms far more primitive, ultimately perhaps from just one or a very few original ancestors. He also proposed a mechanism, natural selection—or, as it came to be called, the survival of the fittest.

The *Origin* was one of the most significant and controversial works of the age—of any age—most particularly because the book was seen to challenge long-held views about religion, specifically the Christian religion and its claims about creation and about the nature of God, of humans, and of our relationship to God. Yet "Dover Beach," with its anguished cry, shows that a loss of faith in the second half of the nineteenth century cannot simply be reduced to the effects of Charles Darwin's great work. Irrespective of science, religion itself was in crisis, searching for some meaning to earthly existence. For a full understanding of the Darwinian story, we must put evolution in context, particularly the context of religion itself, before and after the appearance of the *Origin.* The aim of this book is to provide precisely such an account.

What I offer is not simply a history of evolutionary thinking set against the trials of religion through the past three hundred or so years, though this in itself is a fascinating story, and I will certainly offer more than just the bare details. By trade I am first and foremost a philosopher, and I have rather different fish to fry. I am interested in the status of evolutionary thinking today, and why so many people— why so many people of a certain religious bent—continue to find evolutionary thinking offensive and threatening. Why is it that we have today two violently opposing camps?

Why, in one corner, do we have the evolutionists, particularly those who support some version of contemporary Darwinism, working away in full confidence that they are engaged in a forward-looking area of science? Feeling, indeed, that each day brings new triumphs of discovery and understanding. Why, in the other corner, do we have evangelical Christians who insist on a literal reading of the Bible and on Genesis as a valid guide to world history? Why do the so-called intelligent design theorists affiliated with this group—men with enviable academic qualifications—argue so strenuously that modern science is on an altogether mistaken track? Why do so many people want to fight Darwinism tooth and nail? How can a theory that seems to one group of people so well confirmed as to be almost self-evident be so unacceptable—indeed, blasphemous—to another group of people? How can this other group of people embrace enthusiastically arguments that seem to the first group to be so obviously false—indeed, ignorant?

I am a very publicly committed evolutionist, an ardent Darwinian, and for three decades now I have been fighting creationists on the podium, in print, on television and radio, in court. Let me reassure, or disappoint, my readers by saying I am not now about to confess to a conversion. In fact, so strong is my conviction that the evolutionists are right and the creationists are wrong about the origins of life's diversity that I am going to take that as given in this discussion. I and others have made this case many times in the past, and I do not want to use up space simply going over old arguments. But I do have something new to offer, I hope, even to those with whom I disagree vehemently. Whether or not they accept organic evolution, all participants in the debate accept as fact that ideas have a history and that understanding the past can lead to understanding the present.

For this reason, I will tell a tale of evolution and religion that will hold surprises for *all* of today's controversialists. The full story is far more complex than any of us, including (especially) us evolutionists, have realized. The dispute, as we shall see, is more than merely a matter of right and wrong. At some deeper level, it involves commitments about the nature of reality and the status and obligations of humans in this reality. In particular, I argue that in both evolution and creation we have rival religious responses to a crisis of faith—rival stories of origins, rival judgments about the meaning of human life, rival sets of moral dictates, and above all what theologians call rival eschatologies—pictures of the future and of what lies ahead for humankind. But these rivals are blood relatives. And, paradoxically, the bitterness of the controversy can be traced in large part to the fact that this is a family feud. The two sides share a common set of questions and, in important respects, common solutions. With these things uncovered, I will offer what I hope are fertile and constructive suggestions for moving forward.

Two clarifying comments. First, a matter of language. The notions of *science* and *religion* are probably the two most important in this book, especially since much of the discussion is about the extent to which these terms can be stretched and applied to areas they do not obviously cover. Because of the complexity of the issues, I shall not attempt a preliminary analysis here, but rather let the usage of the terms emerge in the course of historical exegesis. I do promise that this problem will not be left to slide and that due care will be given to meanings. In the case of science, I will be particularly keen to give the criteria for good or mature or what I shall call professional science, and to distinguish this science from rivals and pretenders like pseudoscience and even popular science. In the case of religion, I will be particularly keen to see in what way one can appropriately talk of an essentially secular area of inquiry as being in some sense religious. I certainly do not want to claim that just because one is keen on something, this qualifies it as a religion.

More immediately pressing is the use of the terms "evolution" and "creation," which take on their present meaning only in the course of my story. The term evolution came to general use in the modern sense in the second half of the nineteenth century, and creation likewise means things today that were not so in the past. Hence, strictly speaking, using the terms in a generic sense courts anachronism, and for that I apologize right now, if not very sincerely. So long as care is taken, however, there need not be distortion. More troublesome is ambiguity. The terms evolution and creation each have at least three meanings. Evolution can mean the *fact* of evolution—the development of organisms in a lawbound way from forms very different, from the earliest organisms to the present. It can also mean the *theory* of evolution—usually, one includes here mechanisms based on Darwinian natural selection. Or the term can mean the whole metaphysical or ideological picture built around or on evolution–strictly speaking, this is called *evolutionism.*

Creation can simply mean the origins of organisms (or anything else, for that matter). It can also be the Judeo-Christian concept of creation by God from nothing—some people would designate this Creation with a capital C. Or it can be *creationism,* the specific, biblically based religion of many (especially American) evangelicals—six days of creation, miracles needed to make species, humans given form last, universal flood, and so forth.

As a matter of fact, although we shall need to tread carefully here and note qualifications, creationists tend not to be that keen on any of the serious forms of evolution, and they have no use for evolutionism. Evolutionists, for their part, are against creationism, and many are not all that keen on involving God in creation (that is, in origins) in any way. When ambiguity is no great danger, or when I am referring to several or all of the various meanings, I will without qualification speak simply of evolution and creation (as in the title to this book). But I do not want to let ambiguity conceal the radical nature of my overall thesis. At the most basic level, the clash is between those who push some form of evolutionism and those who push some form of creationism—a clash between two rival metaphysical world pictures. However, I shall argue that the very idea of evolution in some way threatens not Christianity (many Darwinians are Christians) but the idea of creationism and perhaps other notions of creation—and therein lies the problem. In other words, there is more to the story of creationism than a simple misunderstanding on the part of religious enthusiasts of the role and nature of disinterested, objective science.

Second, a matter of intention. By now it must be clear that my text is not a history just for history's sake. It has a purpose: to show where we are today and why. I do not pretend to give an equally balanced history of science and religion but rather to give a history of evolution, with a special focus on how Christianity impinges on that story. Moreover, because the clash between evolution and creation (and more especially creationism) is, today, essentially American, the emphasis will be on elements leading up to this uniquely American controversy. In the earlier chapters, I include discussion of ideas and movements on the continent, but generally the spotlight is on Britain and then, as the action moves west, on the New World. Also, because I offer a history with a purpose, the coverage focuses on issues pertinent to today's disagreements. Important topics like the details of human evolution are touched on only when they are germane to the central discussion.

Always heard in the background, and giving continuity to the story, is the "eternal note of sadness" about which Arnold wrote—the loss of faith. The perceived failure of conventional religion to continue in the role it had played for nearly two thousand years of human history infects, depresses, and inspires all the actors who march across the "darkling plain" where the events of my story unfold.

I

Christianity and Its Discontents

When I survey the wondrous Cross
on which the Prince of Glory died,
my richest gain I count but loss,
and pour contempt on all my pride.

Forbid it, Lord, that I should boast,
save in the cross of Christ my God:
all the vain things that charm me most,
I sacrifice them to his blood.

See, from his head, his hands, his feet,
sorrow and love flow mingled down!
Did e'er such love and sorrow meet,
or thorns compose so rich a crown?

*M*uch Protestant theology is packed into the beginning lines of this beloved hymn, published in 1707 by the English Congregationalist minister Isaac Watts. The focus is on Jesus, who freely accepts sacrifice on the Cross for our sakes, out of pure love. We are sinners, and he has put himself in the way of our deserved punishment, in an act of substitutionary atonement. The hymn stresses our insignificance, our worthlessness beside our lord, and the futility of believing that we can do anything, save through him. Yet, through the unmerited grace of God we have the possibility of eternal salvation—if, and only if, we approach him through faith and commit our lives to Christ. In its concluding lines the hymn stresses our obligation:

> Were the whole realm of nature mine,
> that were a present far too small;
> love so amazing, so divine,
> demands my soul, my life, my all.

The debt of grace must be paid, but it can be paid only in the currency of unerring, total faith and commitment. This exchange—at the very root of Christian salvation—is a crucial insight into the religious mind that we must hold onto throughout this book.[1]

To begin our story, let us go back nearly two thousand years in time and travel to one of the colonies on the outskirts of the Roman Empire. One doubts that Jesus saw himself as the founder of a world religion. On that bitter Friday afternoon when he was taken down from the Cross, there was no belief system that could be called Christianity. Nor on the Sunday following or in the weeks after, when he visited with his disciples before he left them for the last time on this earth. No dogma, no set of rules for everyday living, no structure, no organizers, no nothing. It fell to the followers, Peter especially and then above all the Apostle Paul, to begin the job of building a religion that could function and endure and give meaning to the lives of Christ's followers. And this task continued for several centuries as the theologians and thinkers (the Church fathers) fleshed out the meaning of Christ's coming—his life, his teachings, and his death—and put it all together in a coherent system.[2]

One of the most pressing problems for those early Christians was their relationship to the Jews. Jesus and his immediate followers, not to mention Paul, were all Jews, and their thinking developed within that context. But Christianity became a religion of people who were not Jewish—people whom the Jews called gentiles. Most Jews rejected the Christian faith. What then should be the attitude of the new Church toward the sacred writings of the Jews? The books of the

New Testament, written by the Apostles themselves or their close disciples in the first hundred years after Christ's birth, became part of the Christian canon after a period of sifting and sorting by the Church fathers. But what about the books of the Hebrew Bible that Christians now call the Old Testament? Were these to be binding upon or relevant to Christians? Some sects, the Manicheans, for instance, thought not and rejected them.

But there were good reasons why Christians should take in the Old Testament, quite apart from the historical context into which Jesus' life and death were set. With the stories of the Pentateuch (the first five books of the Old Testament) particularly, one could start to make full sense of Jesus' appearance on earth and explain why his dying on the Cross was relevant to the fate of humans. The story of Adam and Eve showed that we are all tainted by original sin, and hence the Redeemer had to die to save us not just from our sinful actions but from our sinful natures. Only through his sacrifice, the necessity of which was made clear in Genesis, could we hope for eternal life.

But the Old Testament offered something more: a historical perspective on the entire world and its living inhabitants, particularly its human inhabitants. It told a story of origins, how the universe was created and began, with a focus on the absolutely central role of humans in the past, present, and hoped-for future of the world. In the Old Testament narrative we have something dramatically new. The Greeks, the great thinkers of the ancient world, left a metaphysics of eternal existence. Change occurred down here on earth, but the universe as a whole essentially had always been here and enjoyed the prospect of an infinite—if not necessarily human-centered—future. The god of the great Greek philosopher Aristotle spent his time forever contemplating his own perfection and was indifferent to the fate of the inhabitants of our globe. The Christian world picture, by con-

trast, was one of beginnings, of growth and development. A picture
where we humans were the main event.

But why not some kind of evolutionary picture? Why not a world
in which organisms grew from minute points, slowly over time? There
were two major reasons. First, early Christians already had the story
of Genesis, which told of God's miraculous creation of the living
world in six days. One can read this tale in many ways, but an evolutionary reading does not come first to mind. This does not mean that
the early Christians were creationists in a modern American sense.
They accepted the stories of Genesis as the correct account, but primarily because they could see no reasons to do otherwise.

When he was young, Saint Augustine, the greatest of all of the
early theologians, had been a Manichean. He knew all of the problems with the Old Testament (the conflicts between the two different
creation stories in the first two chapters of Genesis, for instance), and
he warned that the very last thing Christianity needed was slavish adherence to the letter of Mosaic text. He stressed that those early stories were written for primitive, nomadic peoples, not for sophisticated
citizens of the Roman Empire. If one has reason, scientific reason, to
reject a literal reading of the Bible and to accept a metaphorical interpretation, then so be it. As it happens, no one at that time thought
that such reason existed, but the possibility was there.[3]

A religion based on faith in the unseen, particularly on the belief
that the Bible is the divinely inspired word of God, is known as "revealed religion." But thinking people in Saint Augustine's time did not
have to resort to revealed religion to justify their faith in creation.
Natural religion—religion based on observations of design in the
world around them, particularly in organisms—also preempted any
kind of protoevolutionary thinking. As early Christians articulated
and developed their faith, they drew on a legacy handed down from
the Greeks which rejected organic developmentalism.[4] Both Plato and

Aristotle had noted that living things seem put together for specific ends, namely, the good of the organisms themselves. To understand how that came about, one must go beyond proximate, or nearby, causes and appeal to "final," or teleological, causes.

Granted, there were a few pre-Socratics who supposed that a kind of protoevolution took place. Empedocles and the atomists thought that pieces of body cohered by chance and eventually became fully functioning organisms. But the greatest Greek philosophers and their followers, including the physician Galen, thought this a false philosophy. Not even infinite space and time would yield complete functioning organisms. One must suppose some principle of ordering, some kind of intelligence at work in creating life. The organic world had a designer—it could not have come about through blind adherence to physical laws. Such was the legacy of the Greeks, and the Christian world bought into it completely.

If the Old Testament provided a story of origins that fit the worldview of early Christians, the New Testament offered a glimpse of the future—eternal salvation for sheep and eternal damnation for goats. Jesus' lifespan came at a time when the Jews were much oppressed by their Roman overlords. Trying to make sense of such trials for a people who thought of themselves as being favored of God, many turned to apocalyptic writings and dreams which envisaged future battles between the forces of good and evil, with the former eventually triumphing.[5] The Old Testament book of Daniel is just such a work of this kind. Not only was Jesus himself influenced by such fantasies, but after his death his followers continued in this pattern. Most influential was the apocalyptic dream of John (not to be confused with the Apostle who wrote the Gospel of St. John), whose Revelation spoke of future travails and clashes. A monstrous battle, Armageddon, would be succeeded by a thousand-year period—the millennium—when Jesus would rule on earth. More conflict would follow, and then good would triumph during the Last Judgment. To early Christians, even more persecuted than Jews, tales like this gave great comfort and meaning to an otherwise dreadful existence.

But as the centuries rolled by and as Christianity rose in power and popularity, people felt less need for apocalyptic predictions. Saint Augustine, in particular, although not excluding Revelation from the canon, dealt with such stories in a highly metaphorical fashion. We should not think of the millennium as an actual time to come, he said, but as an event already arrived with the incarnation of Christ himself. He urged those who spent time trying to calculate the date of the end of the world to "relax your fingers and give them a rest." And this set the pattern for centuries. Toward the end of the twelfth century the Cistercian monk Joachim of Flora revived a more literal interpretation, with a three-part, upwardly rising interpretation of history, from the Age of the Father, through the Age of the Son, to the future Age of the Spirit. But Aquinas and others had little time for this thinking, and through the medieval period a more static and stable world picture prevailed.[6]

The Crisis of Faith

The Reformation was and was not *the* major crisis in the history of Western Christianity. It broke the power of Rome, by dividing Christians into Catholics and Protestants, but it did not at once lead to unbelief. Indeed, in their own way, early Protestants were even more ardent in their commitments than were Catholics, especially in their emphasis on the significance of the Bible in reenforcing each person's faith. Among Catholics, faith was expressed through adherence to the tradition and authority of the Church, rather than through reading and following the letter of the Scriptures. Protestants, by contrast, often consciously reached back to the theologies of the early Church fathers for the precepts of their faith. Sin, sacrifice, substitution, salvation—these Reformation ideas were taken right out of Augustine. Even some of the grimmer and odder views of the reformers—for instance, John Calvin's endorsement of predestination, which held that God alone chooses who shall be saved and who condemned—had early Christian roots. But the Reformation did have repercussions and implications that slowly began to undermine the very foundations of Christianity. People began to entertain the appalling possibility that all religious teaching might not be true, that religion itself might be false. Why did this worry, this doubt, start to appear?

After the initial burst of enthusiasm for Church reform in the sixteenth century, the implications of religious differences started to sink in. Is God a Catholic? If not, then he must be a Protestant. But if so, what kind of Protestant—Lutheran, Calvinist, Anglican, or some other? What if he is none of the above? By the end of the seventeenth century, people were becoming aware of other civilizations. Why should these folk be regarded as mere savages—"pore benighted 'eathen," as Rudyard Kipling would later put it sardonically? These peoples had religions totally alien to Christianity. Growing familiarity with India—a country with a sophisticated and venerable religious heritage—caused a major cultural disjuncture. What should one make of these people? Could one perhaps find some stripped-down religious core that included their beliefs along with Christianity, and indeed with Judaism, Islam, and other great religions? With the building of empires, the question of religious tolerance became a pressing practical matter, for the British particularly were disinclined to push their own religious views on other cultures. Such practices would only foment discontent with the masses and would most certainly be bad for trade.

Edward, Lord Herbert of Cherbury, who even in the first half of the seventeenth century was worrying about the diverse claims made in the name of the Lord, posed this theoretical question: "What shall the layman, encompassed by the terrors of diverse churches militant throughout the world, decide as to the best religion? For there is no church that does not breathe threats, none almost that does not deny the possibility of salvation outside its own pale."[7] Destined to become known as the father of English deism, Herbert worked toward the idea of a religion that postulated a God and a need to worship him and the obligation to be moral, but little more. As deism developed, increasingly it severed its Christian roots and became a belief in an Unmoved Mover, who designed and set the universe in motion and then stood back as his handiwork operated through unbroken physical law.

Deism fit well with the growing triumphs of science, which in the sixteenth century were starting to pressure the authority of the old traditions. One could certainly be a Christian and a Copernican, but only the naive could pretend that no shock occurred. What did a Copernican make of the claims in the Old Testament that the sun stopped for Joshua, for instance? Far worse was the way heliocentrism displaced and downplayed the significance of the earth in the overall scheme of things. No longer were we humans at the center of things, but off to one side, revolving around something bigger and brighter. And the phenomenon of stellar parallax, or rather of its absence— that is, the failure of stars to show displacement as the earth moved through the heavens—meant that the universe, with its myriad stars, had to be simply huge. Back in the tenth century, the Arabs had worked out that the universe was about 98 million miles across. Thanks to Copernicus, it appeared that, at a minimum, the universe was 400,000 times bigger than this. The idea of an infinite universe, with millions of massive stars equivalent to our sun, was not far off, and what intellectual price would a Christian have to pay then to maintain the unique status of our globe? Suddenly all of those Genesis stories about fashioning humans from dust in the image of God started to seem more a function of ignorance than of reason.[8]

The philosophers of the seventeenth century were not helping greatly either. Thomas Hobbes suggested that the natural state of humanity is full of conflict and strife ("brutish and short"), with no need to invoke original sin to explain our predicament. John Locke proved to the satisfaction of many that there are no innate ideas, citing as evidence all the tales about diverse moral and social practices around the globe. Above all, there was René Descartes, who may have remained a good Catholic but who certainly gave food for thought to more than just Catholics through his skeptical philosophy. In the *Meditations,* he introduced his method of systematic doubt, which undercut not only the evidences of the senses but even the existence of a benevolent god. Could it not be that some all-powerful, evil demon is playing with me and deceiving me at the highest or deepest levels, he asked? Descartes himself claimed to rescue the situation with his discovery of an indubitable premise, *Cogito ergo sum*—I think, therefore I am—but many were unconvinced that, having dug himself into a hole, he was now clever enough to escape.[9]

With these philosophical influences came, at first slowly, a willingness to look at religion in the light of reason, and especially to look at the sacred writings as if they were human-produced rather than simply transcriptions of divine dictation. And what this practice—later to be known as "higher criticism"—showed was that major problems of scholarship had to be tackled before one could attempt a fruitful reading of the Bible. The book was a very human document with many authors, and not all of them consistent one with another.

Along with revealed religion, natural theology was also under pressure. If the world is such a wonderful example of design, why do we have so much misery and pain? The philosopher Gottfried Wilhelm Leibniz had responded that even God cannot go against necessity, and that although (being all-good) he created the best of all possible worlds, this did not mean he could eliminate creation's downside. Fire burns, and the result can be a good thing; but fire can also kill, and that is a bad thing. On balance, however, better to have fire than not. This kind of argument was parodied by the French writer Voltaire in *Candide,* where he invented the Leibnizian philosopher Dr. Pangloss. Whenever anything goes wrong—and much does—Pangloss sees it as the "best of all possible things in the best of all possible worlds." A laughably ludicrous conclusion—then as now, humor trumped any amount of serious philosophical argument.

Gradually at first but with increasing speed, new technologies brought social and cultural changes to Europe, especially to Britain, that undermined religious certainty. As people started to leave the countryside and take jobs in towns and cities, the hierarchical and settled state of a rural, farming society started to change, and with it the dominant religious ideologies. Thanking God at the harvest festival made little sense to a factory worker in Manchester or a collier in Durham. One's foreman and the boss were bigger and more immediate authority figures than the squire and the parson. To many, what may have worked in the past no longer seemed relevant to the present, and even less so to the future.[10]

Not all of these changes came at once or together. And deism, while it is not Christian, is not atheism, nor is it necessarily the first step on the road to atheism. But the changes were starting to corrode the old confidence—shared by Catholics and Protestants alike—that God created heaven and earth, that he sent his son to die for our sins because he loves us, that he interferes miraculously in his creation, and that we must submit to his will.

What was to be done? Some—in the manner of King Canute—tried to hold back the tide. They refurbished revealed religion, and they polished up natural religion. This was very much the mode of the established Church of England, which had a long tradition of trying to stay afloat by giving a little and resisting a little more. Its American branch—known as the Episcopal Church—would do likewise. At the end of the eighteenth century, drawing on two hundred years of theological argumentation in his *Evidences of Christianity* (1794), Archdeacon William Paley had staked his case for the authenticity of the Gospels on the willingness of the disciples to suffer death for their faith in the divine nature of Jesus. In 1802 he switched his emphasis away from revealed theology, arguing in his *Natural Theology* that modern science and reason likewise confirm the existence of the Christian deity. Famous even today is Paley's argument that, as the telescope has a telescope maker, so likewise the eye has an eye maker—the great optician in the sky. But for many this kind of argumentation no longer convinced, or at least it no longer convinced just on its own. They wanted something more or something different. Two major responses, at different ends of the spectrum, stand out.

The first reaction to the eighteenth century's crisis of faith was simply to opt out of the conflict. In this view, reason and evidence, made supreme, are tools of the devil, and on the really crucial issues they are deceptive. The way to God is through an open and loving heart, through emotional commitment, not rational choice—that is, through faith and conversion. Thus, the Protestantism of the Reformation gave way to evangelicalism, which acknowledged God as the lord and master who gave his life freely on the Cross for our sins and who now demands of us obedience to his word. The God of evangelicalism is the engaged God of the theist, not the noninterfering god of the deist. He is the God who not only created heaven and earth but manipulates his creation as needed, and without whom his creatures, down to the smallest sparrow, are absolutely helpless.

Most famous and important of the eighteenth-century evangelists were the Wesley brothers, John and Charles, the somewhat inadvertent founders of Methodism. In 1738, at Pentecost, John, an Anglican clergyman and graduate of Oxford University, underwent a profound religious experience: "I felt my heart strangely warmed. I felt I did trust in Christ, Christ alone, for salvation; and an assurance was given me, that he had taken away *my* sins, even *mine,* and saved *me* from the law of sin and death."[11]

This set the pattern for a major shift toward emotional and heart-stirring preaching and singing—particularly in the open air and particularly among those excluded from the upper levels of the social order. Not by chance did Charles Wesley become one of the greatest hymn writers of all time. To be saved (converted), one had to confess one's sins and accept the freely given grace of Jesus Christ. Efforts to earn salvation counted for naught, a sentiment at the heart of the oft-sung hymn, "Rock of Ages," written by Augustus Montague Toplady in 1776:

> Not the labor of my hands
>   Can fulfill Thy law's demands;
> Could my zeal no respite know,
>   Could my tears forever flow,
> All for sin could not atone;
>   Thou must save, and Thou alone.

Still, Methodism broke from the strict Calvinist doctrine where salvation came purely by the grace of God and was predestined. It moved instead toward a position known as Arminianism, where the willingness of the individual to make a leap of faith determined whether he would be saved. After conversion, to demonstrate his faith, a convert was expected to evangelize among others and also to perform good works—an outward sign of an inward state of grace, but not the price of admission to heaven, which Jesus had paid on the Cross.

Although the Wesleys and some of the other Methodist leaders, being educated men, were by no means averse to science and manufacturing—indeed their Gospel went down well in the industrial north of England—they emphasized a simple faith, backed by diligent reading of the Bible. Even some women became preachers, although, as George Eliot noted wryly in *Adam Bede*, not everyone was happy with this development. By the end of the eighteenth century, 70,000 people in Britain identified themselves as Methodists. And this does not count the evangelicals who belonged to other denominations or remained in the "low church" wing of Anglicanism.

America

Then and now, religion was a bigger factor in daily life and thought in North America than in England.[12] The British colonies had been founded by people in whom religious beliefs and yearnings ran deep. In the stress and strain of carving out a new life in a new world, religion continued to sustain what they saw as a divine mission. This mandate was reinforced by the first Great Awakening, a communicable conversion experience that raced through the colonies, bringing thousands to a sense of their personal sin, their need for repentance, and their obligation to spread the Good News. Between 1739 and 1742 the flames of revival were fanned by the preaching tour of the itinerant English Methodist preacher George Whitefield.

Going back as far as the colonists' escape from Britain in the early seventeenth century, the dominant American theology, especially in New England, had been more strictly Calvinist than middle-of-the-road Anglican. From a Calvinist perspective, an angry God made the decision about who would be saved, and sinners themselves could do nothing to earn salvation. As the eighteenth century took its course, Calvinism was moderated and the individual's response to God's offer of salvation was acknowledged as critical. By the end of the century there were almost as many Methodists in the New World (60,000) as in the Old.[13]

Nowhere was the evangelical mind of Americans better revealed than in the thinking of the greatest religious genius of his age, the New England pastor Jonathan Edwards.[14] Although theologically he was a very old-fashioned Calvinist indeed, Edwards encouraged the Awakening and was at the forefront of those who saw the significance of a one-on-one encounter with God at the moment of conversion, without need or possibility of rational argument. The title of one of Edwards's well-known sermons, "A divine and supernatural light, immediately imparted to the soul by the spirit of God, shown to be both scriptural and rational doctrine" (1734), stresses the fact that real contact with the deity consists of emotion and faith rather than reason. Edwards insisted that faith in Jesus Christ as the son of God does not come through our minds or senses, in the way that scientific knowledge is obtained. "He [God] imparts this knowledge immediately, not making use of any intermediate natural causes, as he does with other knowledge." Edwards used an analogy:

> It is out of reason's province to perceive the beauty or loveliness of any thing: such a perception does not belong to that faculty. Reason's work is to perceive truth and not excellency. It is not ratiocination that gives men the perception of the beauty and amiableness of a countenance, though it may be many ways indirectly an advantage to it; yet it is no more reason that immediately perceives it, than it is reason that perceives the sweetness of honey: it depends on the sense of the heart.—Reason may determine that a countenance is beautiful to others, it may determine that honey is sweet to others; but it will never give me a perception of its sweetness.

So with our apprehension of God, his son, our fallen nature, and his infinite love and grace toward us.

But Edwards did not leave things there. Edwards, like all other Protestant Christians of his age, demanded that religious experience be confirmed by the Scripture. And so he quoted appropriate passages of the New Testament. "For God, who commanded the light to shine out of darkness, hath shined in our hearts, to give the light of the knowledge of the glory of God, in the face of Jesus Christ" (2 Corinthians 4:6). Edwards and his fellow Christians were biblical literalists insofar as they believed in divine creation, but they were not, in any recognizably twentieth-century sense, creationists. They certainly read the Bible as the true Word of God, but they recognized that all reading, especially of a divinely inspired text, demands interpretation, and eighteenth-century believers were as much given to exegesis as those of any other age.

Interpretation became inextricably linked with millennial thinking among Protestants. At first the reformers were no more enthusiastic about apocalyptic speculations than were philosophers of the Catholic Church. But partly because Protestants put so much emphasis on the significance of the Bible and partly because the break with Rome brought conflict and persecution, apocalyptic thinking revived during the Reformation. Luther himself was a cameo for this development. At first the Book of Revelation had no place in his Christ-centered theology, but later he started to speculate as fancifully as any early Christian. And with such speculation came interpretation. Lined up against the forces of good in Revelation was an array of monsters and devils—Satan, the Antichrist, serpents and many-headed dragons, the Whore of Babylon, the fearsome Gog from the land of Magog, and more. Few if any took these beasts literally but rather spent many happy hours trying to identify their true identities: is the Antichrist the Catholic Church and the Whore of Babylon the Pope, or is the

Pope himself the Antichrist? One persistent line of thought saw the real threat of the latter days as coming from farther east—from Arab and Muslim lands.

Such millennial speculations found a ready home in the English-speaking world, first among Elizabethans, who felt threatened by Catholic Europe, and then, in the seventeenth century, among the embattled Calvinists, who wanted a sterner, less tolerant religion than the Anglican Church. Apocalyptic thinking made a natural transition to the New World with the Pilgrims and other religiously persecuted emigrants, as the most sober of thinkers tried to tie in the dreams of Revelation with the harsh realities of life in America. The influential New England divine Cotton Mather was forever speculating on the date of the end of time. Disappointed by 1697, he pushed the time forward to 1716, and when that failed he moved dates onward yet again.

Jonathan Edwards fit comfortably into this tradition, spending much time on the prophecies of Daniel and Revelation, trying to interpret current events, such as various battles between the English and French, in the light of past predictions, and very naturally using metaphor and analogy as needed. For instance, Revelation talks of a great hail falling from heaven, "every stone about the weight of a talent." Edwards did not take this literally but rather as a sign of the "strong reasons and forcible arguments and demonstrations" that will batter to pieces "the Kingdom of the anti-Christ," that is popery, "as if they were dashed to pieces by stones from heaven."[15]

Progress

Edwards did not exclude reason from every significant role in Christianity. Reason may not have been all-important, it may have played no part in the greatest insights that we humans receive, but it too was a gift from God, he believed, and had a place on the human stage. As the title of his 1743 sermon made clear, the "divine and supernatural light" of salvation must be supported by "rational doctrine." Edwards spent a considerable portion of his sermon explaining why it is reasonable to expect God to speak directly to us. As the analogy with sweetness underlined, the apprehension of God and his divine messages is simply not the sort of thing that human reason could handle—and reason alone convinces us of that. "It is therefore congruous and fit, that when it is given of God, it should be nextly from himself, according to his own sovereign will." As the Wesleys had preached in Britain, the evangelical response did not reject reason—or the modern science of the day, for that matter—but rather construed the essential connection with God as being beyond the reach of reason.

But what of those at the other end of the spectrum, who wanted to give reason a more central, up-front role in religious life? Their response to the crisis of faith centered on the notion of progress, the idea that humans are gradually improving their lot, socially, intellectually, morally.[16] They started with the assumption that early societies were simple and savage and that through time humankind had progressed, leading to the civilized, functioning nations of the present day (at least in the West). And great though the advances had already been, the future promised to be even brighter, as human civilizations climbed upward toward the light. Where progress distinguished itself from other thought systems was in stressing that advance comes through human effort—we make the difference and we are responsible for change, especially change for the better. Progress could become a plausible idea only after the Scientific Revolution, whose great advances and insights convinced many people that we humans could now foresee the future and control our own fate.

It would be easy to conclude that progress was at sharp odds with Christianity, and this seems true for a Christianity that put all hopes of a better state entirely within the hands of Providence. Such was the position of Saint Augustine, who saw as many evil things being created by humans as good things and who therefore concluded that on this earth we will always have a battle or balance between the sin of humankind and the good given by the Creator.[17] But more accurately one should say that progress was at sharp odds with any ideology, Christian or otherwise, that denied human autonomy and our ability to work things for the better. This included Protestant sects, such as Arminianism, which gave humans the power to choose or reject Christ but still left the real work to the Almighty, and Catholicism, which gave the Church alone the power (and understanding) necessary to change things for the better. Conversely, a religious system—even a Christian one—that was willing to assign humans some part of the work could be construed as progressive, even if it did not grant everything that a nonbeliever might demand. In a sense, therefore, rather than thinking of progress as an alternative to conventional religion in the late eighteenth century, it is more accurate to think of progress as a world system that was trying to challenge or improve on older world systems, especially traditional Christianity.

In France particularly there were outright atheists for whom the doctrine of progress combined with a kind of materialism was in itself enough to explain human existence.[18] Even for those who did not want to go that far, the very point of progress was to oppose Christianity directly, not just its theological doctrines but also the social structure and status of the Church, which supported the *ancien régime*. To quote the greatest enthusiast for progress, the Marquis de Condorcet: "Contempt of human sciences was one of the first features of Christianity. It had to avenge itself for the outrages of philosophy; it feared that spirit of investigation and doubt, that confidence of man in his own reason, the pest alike of all religious creeds."[19] The *philosophes* (as the French thinkers were called) were working and writing in a repressive society, controlled by monarch and Church and other established institutions. It all eventually flew apart during the Revolution.

Britain was a very different sort of society. It had gotten revolution out of its system in the seventeenth century. If the result was not much of a democracy by our standards, at least the nation had a functioning parliament, and those who were skilled, talented, prudent, and adventuresome had a good chance to succeed. This happened most particularly in industry, for the British were the first to harness nature to massive schemes for producing goods mechanically rather than by hand. In ironworking, pottery, wool, and above all cotton, the British introduced factories and ways of turning power—especially the power of coal—to good account. This led to views of progress much more in tune with the pragmatic values of the day—utilitarian values that favored doing things efficiently, through division of labor, to increase happiness and decrease misery.

Although the British progressionists were often not particularly enthused by the Anglican Church, they had no underlying hostility to theological commitment, so long as ideology was well-integrated with progressive social goals. The economist Adam Smith, when speaking of the power of self-interest to organize society for the greater good, invoked the metaphor of an "Invisible Hand." A repeated analogy of the time was that a deity who works through the laws of physics is far superior to one who works through miracles, in the same way that an industrialist who works through the laws of physics is far superior to a cottage laborer who works with his hands.

Strange as it may seem to us today, apocalyptic thinking stood behind progressionist philosophy no less than the philosophy of faith. One of the progressionists' favorite biblical passages was Daniel 12:4, which states that in the latter days of earth's history "many shall run to and fro, and knowledge shall increase." As one writer put it, "the *knowledge* that shall be *increased*, according to the prophet Daniel, means nothing less than a fruitful, influencing knowledge, a knowledge of things pertaining to God and true godliness."[20] Many people, including Isaac Newton, who devoted as much energy to biblical interpretation as he did to theoretical physics, took this as an invitation to push our God-given talents to the extreme, in order to complete God's creation. Although divine interventions could still be expected, the tendency was to regard the future as naturalistically as the present: the "Government of the *Millennial Kingdom* will not be altogether different from That of the Ante-millennial or present *Kingdom of Christ.*" An increasingly popular interpretation of Revelation put the coming of Christ at the end of the millennium rather than at the beginning. Our biblically reinforced task was essentially to prepare for the good times to come, rather than to sit obediently and passively until God himself takes charge. A position like this, the variety of millennialism that stresses human effort and puts Christ's coming later, is known as "postmillennialism." The kind that has Christ coming before the millennium is "premillennialism." Someone like Augustine who opts out of the prophecy business is an "amillennialist."[21]

America of course did have its revolution, although (unlike the French Revolution) it was fought against British oppressors by colonists who now considered themselves Americans rather than British subjects. The causes of the revolution were many and complex, but before and after the decisive break, progressive ideas, generally imported from Britain and France, had significant influence. The importance of millennial speculations—Jonathan Edwards was one millennialist among many—is still a matter of much historical debate, but no one denies that they had some impact, maybe even major impact, on revolutionary thought.[22] Certainly, there were lots of not-so-subtle hints about where the thousand-year rule would take place.

> A new Jerusalem sent down from heav'n
> Shall grace our happy earth, perhaps this land,

Whose virgin bosom shall then receive, tho' late,
Myriads of saints with their almighty king,
To live and reign on earth a thousand years
Thence call'd Millennium. Paradise anew
Shall flourish, by no second Adam lost.[23]

As in Britain, in America also one finds ideas about progress connected to deistic beliefs. People in the newly independent nation developed notions of free trade, an end to the poor house, support for the old and the sick, and even a kind of progressive taxation while reading Thomas Paine's *Rights of Man* and the *Age of Reason*. In those tracts Paine attacked the forces of the establishment and especially their ideology. "What is it the Bible teaches us?—rapine, cruelty, and murder. What is it the Testament teaches us?—to believe the Almighty committed debauchery with a woman engaged to be married, and the belief of this debauchery is called faith." Although the founders tended to be more circumspect in their comments than Paine, a kind of progressivist deism marked the thought of Franklin, Washington, and Jefferson, among others. Franklin, who labored his whole life to increase knowledge and learning, gave up the idea of a personal god. George Washington, although a somewhat occasional Episcopalian, never took communion. Thomas Jefferson spoke of the Trinity as the "Abracadabra" of confidence men.[24]

People like this were not atheist or even what came to be known as agnostic. Often they would insist indignantly that they were genuinely Christian. But they were moving beyond a belief in the God of Providence, who decides all and is uniquely responsible for our salvation. As happened in Britain, influential Americans came out of the eighteenth century more on the side of reason than of faith. And this starts to push us toward ideas of evolution.

2

From Progress to Evolution

She invited me to kiss her forehead, cheeks, eyes and mouth, and I obeyed. I don't think there was any harm in that, but her pleasure increased, and as I was only too glad to add to her happiness in any innocent way, I kissed her again on forehead, cheeks, eyes and lips. The hand she had rested on my knee wandered all over my clothing from my feet to my girdle, pressing here and there, and she gasped as she urged me in a strange, low voice to redouble my caresses, which I did. Eventually a moment came, whether of pleasure or of pain I cannot say, when she went as pale as death, closed her eyes, and her whole body tautened violently, her lips were first pressed together and moistened with a sort of foam, then they parted and she seemed to expire with a deep sigh.

*O*ne suspects that someone whose idea of a really good joke was to write a pornographic novel about lesbian nuns was going to have a fairly casual attitude toward the creation stories of Genesis, and so indeed it proved. Denis Diderot (1713–1784), author of *La Religieuse (The Nun)*, was one of the great figures of the French Enlightenment, for many years the editor of the *Encyclopédie*, a wonderful attempt to gather up all human knowledge. An atheist and a progressionist, he was one of the very first whom we can describe properly as an (organic) evolutionist.

No surprise here. The cultural idea of progress led to the biological idea of evolution.[1] People took the cultural notion and read it into the living world, seeing the consequence in terms of development from the primitive up to the complex and valuable—namely, humans. Often, then, they took the biological notion and read it back into culture as confirmation of their beliefs about progress. They moved from progress to evolution, from evolution to evolutionism (often without much worry about causal theories), and then back to progress.

This short-hand analysis sounds somewhat glib and cynical, too easy to be true. So let me stress that the eighteenth century did see much scientific activity, and some was certainly pertinent to the issue of evolution—digging up fossils, looking at geographical distributions, morphological studies, and so forth. But metaphysical ideas played as big if not a bigger role in the origins of evolutionism. Few people became evolutionists and then lost their religious faith as a result. Rather, Matthew-Arnold-like, people became (a bit/moderately/totally) dissatisfied with their religion and then went looking for something else as a supplement or alternative. Evolution was part of the something else. It was attractive not only because it offered an alternate world picture but also because at some level it had the cachet of being scientific rather than religious, precisely the kind of thing that the searcher fervently desired.

Complementing and fleshing out the idea of progress was a theme that goes back to Aristotle—the great chain of being.[2] It claimed that all organisms can be put in a continuous line from the simplest to the most complex, from monad to man, as people often said. A number of thinkers at the beginning of the eighteenth century toyed with various versions of the chain. In itself the chain was not evolutionary—it was more a fixed ladder than a moving escalator—but it was pointed in the right direction, for those who were so inclined.

Diderot was at the front of the pack of the biological progressionists. He speculated that species are born, live, and die after reproduction, just as individual organisms do.

Just as in the animal and vegetable kingdoms, an individual begins, so to speak, grows, subsists, decays and passes away, could it not be the same with the whole species? . . . would not the philosopher, left free to speculate, suspect that animality had from all eternity its particular elements scattered in and mingled with the mass of matter; that it has happened to these elements to reunite, because it was possible for this to be done; that the embryo formed from these elements had passed through an infinity of different organizations and developments . . . that it has perhaps still other developments to undergo, and other increases to take on, which are unknown to us; that it has had or will have a stationary condition; . . . that it will disappear for ever from nature, or rather it will continue to exist in it, but in a form, and with faculties, quite different from those observed in it at this moment of time.

This mirrored exactly his thinking about culture, where societies were born, developed, and finally grew old. Savages were simple; Europeans were complex. "The Tahitian is at a primary stage in the development of the world, the European is at its old age. The interval separating us is greater than that between the new-born child and the decrepit old man."[3]

Other people speculated in similar sorts of ways. One was the man who is rightly considered the father of French evolutionary thinking, Jean Baptiste Chevalier de Lamarck—a biological progressionist all the way: "Ascend from the simplest to the most complex; leave from the simplest animacule and go up along the scale to the animal richest in organization and facilities; conserve everywhere the order of relation in the masses; then you will have hold of the true thread that ties together all of nature's productions, you will have a just idea of her *marche*, and you will be convinced that the simplest of her living productions have successively given rise to all the others."[4]

More of a deist than an atheist, Lamarck argued that organisms respond to "needs" and that this brings on change. He made no secret of the fact that he was a social and cultural progressionist. He was, after all, a nobleman who did well during the Revolution, although he did find it politic to change his name from the Chevalier de la Marck to plain Citizen Lamarck. He was a good friend of the *philosophes*, and much of his major evolutionary work, the *Philosophie Zoologique* (1809), was taken up with progress-affirming speculations. In later writings the same theme emerged. By 1820 Lamarck was running everything together, as he argued that humans have risen from a primitive state through their own efforts, and that Europeans now represent the most perfect form of being. Lamarck even went so far as to argue that, ideally, society would be run by a few philosopher-kings and that science and scientists would figure high in such a scheme. Again, no surprise here.

Cultural change of course involves learning or discovering new things and then passing them on to the next generation. Lamarck took this "inheritance of acquired characteristics" and read it back into the biological world as a mechanism for producing permanent change in species. (To take a simplistic example, the giraffe who stretches his neck to reach the choicest leaves will pass this trait along to his offspring.) This process today is known as Lamarckism, although the idea certainly did not originate with him, nor was it the chief form of change that Lamarck proposed. The real basis of biological change, he believed, was progress up the ladder of life. The inheritance of acquired characteristics was a kind of secondary process that could explain some variations and diversions.

Similar evolutionary thinking was found across the channel in Britain. From around the time of the American Revolution, Erasmus Darwin, the physician grandfather of Charles Darwin, became impressed by fossil discoveries thrown up during excavations for a new canal, and this inclined him toward evolution. He was also an arch-progressionist (as well as a poet), who endorsed a tree of life that, as Darwin himself put it, went from monarch (butterfly) to monarch (king).

> Imperious man, who rules the bestial crowd,
> Of language, reason, and reflection proud,
> With brow erect who scorns this earthy sod,
> And styles himself the image of his God;
> Arose from rudiments of form and sense,
> An embryon point, or microscopic ens![5]

Erasmus Darwin's notions of evolutionary and cultural progress were inextricably tied together. He was a member of the Lunar Society, a British Midlands group of industrialists and inventors that included the potter Josiah Wedgwood, the ironmaster Matthew Boulton, and the chemist Joseph Priestley. In an effort to be more efficient in their work, these men would get together to discuss ways of applying science to industry. Darwin specialized in agriculture, writing tracts on the subject and coming up with improvements for plows and windmills and other inventions.

Like his fellows, Erasmus Darwin was an ardent utilitarian who wanted to improve the lot of humankind and promote happiness. He was in favor of change, even revolution, and was a good friend of Benjamin Franklin. He praised the rebellion on the other side of the Atlantic, and—until things went dreadfully wrong—he was also well disposed toward revolutionary events across the Channel. He was ever happy to break into verse to celebrate the triumphs of his fellow men of business and industry: "So with strong arm immortal BRINDLEY leads / His long canals, and parts the velvet meade."[6]

Darwin drew an explicit analogy between the progress of culture and the progress of biology, the one feeding into the other and then back again. The idea of progressive evolution in the living world was "analogous to the improving excellence observable in every part of the creation . . . such as the progressive increase of the wisdom and happiness of its inhabitants." All of this was then wrapped up with some grand sentiments about the Creator—for Erasmus Darwin was no atheist or agnostic. Like Lamarck, he firmly believed in the deity. For him, evolution represented the triumph of unbroken law, the apotheosis of God's standing and worth. Everything was planned beforehand and went into effect through the laws of nature. "What a magnificent idea of the infinite power of *The Great Architect! The Cause of Causes! Parent of Parents! Ens Entium!*"[7]

Pseudoscience

Progress—both cultural and biological—was what made evolution attractive. But progress—both cultural and biological—was also what made evolution dangerous. Many saw progress as a radical ideology that challenged conventional Christianity and the social systems with which traditional religion was entwined. Social and religious conservatives wanted no part of it. Erasmus Darwin was made the butt of savage parodies by right-wing politicians. By the end of the eighteenth century, these people were disgusted with or even (with some good reason) terrified by the ideology of progress. Its supporters had sided with the American revolutionaries, and then—at least until things got out of hand—with the French Revolution. First the French hoards went berserk, and now Napoleon was starting to threaten Britain with war and rapine unknown since the disturbances nearly 150 years before. The progressionists had to be stopped.

Not that the critics were motivated purely by political or religious factors. All of the evolutionists' opponents stressed that the sort of stuff the movement was producing and promoting was not real science—not mature science, as one might find in physics and chemistry. It was rather a thing of the popular domain, dressed up to be more than it ought to be. Evolution was an ideology masquerading as science. A pretender. And never just evolution but always evolutionism. The French particularly were sensitive to just those standards we today regard as good science: careful observation, use of mathematics as needed, experimentation, measurement, prediction, sober restraint from wild hypotheses.[8] They were also well aware of the marks of bad science.

By bad science I do not refer to science written for the general public, which cuts down on mathematics and the like to make clear the essential points. At the beginning of the nineteenth century, the great French physicist Laplace was the exemplar of the professional scientist who could also write for the general public. His *Exposition du système du monde* and his *Essai philosophique sur les probabilités* were wonderful nontechnical expositions of his theories of astronomy and of probability respectively, aimed deliberately and explicitly at the nonexpert. These works were then—and they remain—paradigms of a more popular kind of science writing. In speaking of bad science, I am rather referring to ideas and theories which, driven by underlying metaphysical commitments, simply violated or ignored all of the standard methods of good science.

Mesmerism, the claim that the body contains a kind of animal magnetic fluid essential for good health, was a case in point.[9] Alarmed by its popularity—neurotic women were given to mass displays of hysteria under its supposed influence—King Louis XVI set up a commission to study the subject. The group, chaired by Franklin and including the chemist Lavoisier, concluded that mesmerism is truly bogus; unquantified and unquantifiable, it made no controlled experiments and predicted only that which was already known. "Since the imagination is a sufficient cause, the supposition of a magnetic fluid is superfluous."[10]

Mesmerism was not genuine science at all. We might call it "pseudoscience," to distinguish it from the mature or "professional science" of someone like Lavoisier and from the "popular science" of Laplace. Professional science is an attempt to uncover the regularities—the laws—that govern the natural world and to bring them together into a unified system of knowledge. Each branch of science strives to be internally coherent and externally consistent with other areas of science, and above all it values making successful predictions about the future course of natural events, in the real world and in the laboratory. To these ends, scientists tend increasingly to rely on formal tools, especially mathematics, to wring answers to puzzling questions from both natural "experiments," such as volcanic eruptions or the spread of infections, and laboratory experiments, whether with mice or molecules. An important mark of professional science is that it tries to divorce itself from the immediate concerns and values of the society that is producing it. In the memorable words of the philosopher Karl Popper, science is "knowledge without a knower." By this he meant that successful professional science stands alone, its truth separate from the interests or cultural background of the person who produces it or who accepts it.

Popular science is akin to professional science, but in aiming to inform nonprofessionals it does not pretend to completeness of proof or evidence. Pseudoscience cares only or primarily for the social values and commitments of its enthusiasts, and manipulates all else to support these ends.

Obviously, all and any kind of science can be used in support of (or criticism of) religious positions and views, but the further one departs from professional science the easier it is to entwine one's ideas about the workings of the natural world with one's religious commitments and social agenda. Probably one should not say that all pseudosciences are religious, or quasi-religious—and certainly one should not say this of popular science. But given that religion incorporates values (moral dicta such as loving one's neighbor as oneself), pseudoscience can be readily folded into a religious position, or used against the religious position of others. Certainly, contemporaries of the evolutionists regarded the hypotheses about the history of life in precisely this way. With good reason, evolutionary speculations were seen as indifferent to the standards or demands of genuine science and more akin to an ideology or value system. Moreover, these ideas were bound together with deism or, worse, with the ideology of progress, both of which threatened traditional Christianity.

The French Connection

One might sneer at the early evolutionists, but new and disturbing facts were flowing in from the worlds of biology and geology and related sciences. These discoveries, along with successes within the physical sciences, pressured biologists to come up with robust theories that could account adequately for the fossil record, in the same way that gravity had explained the movements of the heavenly bodies. Several respectable hypotheses were offered at the beginning of the nineteenth century. The dominant one came from Georges Cuvier, the father of comparative anatomy and a leading figure in French scientific circles. He was desperate to put biology on a professional basis—to make it stand proudly alongside the physical sciences and eschew the wild evolutionary speculations of Lamarck and his ilk.[11] Partly, this drive came from the purest of scientific motives. Cuvier was a professional biologist, and he wanted his chosen field of study to garner its full measure of status and respect. But partly, this drive was politically motivated. Having been born in one of the French border states adjoining Germany, Cuvier was a Protestant at a time when (first under Napoleon and even more so after the restoration) Catholicism in France was both powerful and conservative. He needed to show that his religion was not a threat to the French establishment, and what better strategy than to portray his own area of expertise, biology, as a mature, objective field of academic study, not at all tainted by crass ideology? Cuvier thus searched for basic principles and methods that would enable him to turn out work that was firmly based in empirical evidence, well integrated into a larger body of fact and theory, and capable of both explanation and prediction.

The legacy of the Greeks came to Cuvier's aid. Organisms are integrated, they function, they work toward ends, they are governed by final causes.

Natural history nevertheless has a rational principle that is exclusive to it and which it employs with great advantage on many occasions; it is the *conditions of existence* or, popularly, *final causes.* As nothing may exist which does not include the conditions which made its existence possible, the different parts of each creature must be coordinated in such a way as to make possible the whole organism, not only in itself but in its relationship to those which surround it, and the analysis of these conditions often leads to general laws as well founded as those of calculation or experiment.[12]

Note the careful mention of laws and experiment and measurement, all the things one associates with professional science. From the conditions of existence, operationalizing it as it were, Cuvier moved to his own domain of interest, morphology. And on the basis of much careful study of many kinds of organisms, he enunciated the secondary principle of the correlation of parts. Every part of an organism must work harmoniously with every other part. "It is in this mutual dependence of the functions and the aid which they reciprocally lend one another that are founded the laws which determine the relations of their organs and which possess a necessity equal to that of metaphysical or mathematical laws."[13]

Cuvier felt that this pointed to another principle, the subordination of parts, and from this one had certain basic groupings *(embranchements)* of animals. He listed four animal *embranchements:* the vertebrates (with backbones), the molluscs, the articulates, and the radiates. "I have found that there exist four principal forms, four general plans, upon which all of the animals seem to have been modelled and whose lesser division, no matter what names naturalists have dignified them with, are only modifications superficially founded on development or on the addition of certain parts, but which in no way change the essence of the plan."[14]

The positive part of his program now expounded, Cuvier could draw the negative conclusion he wanted so much to draw. You cannot go from one *embranchement* to another. Evolution is simply impossible. It cannot have occurred. As empirical evidence, he pointed to the mummified animals brought back from Egypt by Napoleon's savants, and noted that although they are very old they show no sign whatsoever of difference from forms that exist today. As for the fossil record, in which he took an intense interest, Cuvier argued that we never find intermediate forms. Animals and plants are always one type or another, never something in between. The evidence argued strongly against evolution, with its embedded notion that life forms progress over time.

Catastrophism

Cuvier hated the ideology of progress. In major part, the hatred came because he held positions of responsibility in France, and as a servant of the state he loathed all movements that threatened instability. He had lived through the Revolution and knew what ideas like progress could bring on. Paradoxically, however—especially given that Lamarck was indifferent to such issues—Cuvier was the person who started to lay a firm foundation for a progressivist reading of the fossil record. He showed that the organisms that came lower in the strata seem generally to be simpler than the ones higher, and that the latter tend to resemble today's forms more closely. He blended this finding into a picture of world history which stressed the importance of water as the major causative element in the formation of the globe.[15]

Cuvier proposed that, periodically in the history of the world there had been major upheavals—he called them revolutions—that seemed to wipe out all life in a given region. At some point thereafter, life would reappear. Apparently organisms were not created anew (whatever form the creation might have taken) but involved immigrations from other parts of the globe. Even if he had not had personal reasons to be careful on this score, mixing science and religion was not a French trait. So, although he was prepared to use Genesis as a historical document, Cuvier did not interpret the fossil record in biblical terms.

This kind of thinking went down well in England.[16] By the 1820s at Oxford and Cambridge, a group of ordained Anglican clergy (a

necessary condition of holding a post) had become increasingly interested in science. They included the geologists William Buckland at
Oxford and Adam Sedgwick at Cambridge, the botanist John Stevens
Henslow at Cambridge, and the most remarkable man of them all,
William Whewell a former scholarship boy from Lancashire who
started out as a writer of physics textbooks and was elected professor
of mineralogy. As was typical, Whewell set about learning more on
the subject than anyone else had ever known, and then went on to
write extensively on the history and philosophy of science, before becoming professor of moral philosophy and Master of Trinity, one of
the richest and biggest Cambridge colleges.

This group seized with relief on Cuvier's thinking. They loathed
ideas of progress and saw evolution as part and parcel of this dreadful
picture. Adam Sedgwick, writing to Herbert Spencer in 1853, said: "I
am no believer either in moral or social perfectibility and I believe that
all sober experience teaches us that there are conditions both moral
and physical which must entail physical and moral pain so long as the
world lasts." These men cherished the idea that design in the natural
world implied an omnipotent designer, and they saw this "final cause"
as an essential link between their science and their theology. They
wrote extensively on the topic, a task made easier by the Eighth Earl
of Bridgewater, who left a large bequest to sponsor a series of works
on natural theology, to which several made contributions. Whewell
wrote a Bridgewater Treatise, as these works came to be called, on astronomy and general physics, and Buckland wrote on geology.

Most of all, the natural theologians loved Cuvier's geology of revolutions, to which they gave the more apocalyptic name of "catastrophes." One and all agreed that Cuvierian geology was the answer to an
Anglican scientist's dream. These scientists were more than happy to
take the progressive nature of the fossil record as evidence of the un-
furling of God's creation. Demonstrating that just because someone is

a biological progressionist they are not necessarily an evolutionist, the British (probably more than Cuvier, who was discrete on the origins of organisms) saw biological progress as evidence of God's creative hand at work in organic life. Providence rather than cultural progress. Sober science and sober religion in a grand synthesis.

3

Growth of a Pseudoscience

Up the close and down the stair,
In the house with Burke and Hare.
Burke's the butcher, Hare's the thief,
Knox, the boy who buys the beef.

*T*he brilliant pre-Victorian Edinburgh teacher of human anatomy, Robert Knox, fell afoul of the laws of the land restricting his profession. Until 1832 the only legitimate supply of cadavers came from the end of the hangman's rope, and, although Britain had many crimes punishable by death, the supply was never enough to meet the demand. Hence, certain enterprising tradesmen took to hovering around the edges of funerals until all of the mourners had left, at which point they would exhume the dear departed, who the next morning would find themselves participating in an afterlife of which they had never dreamed. Two such "resurrection men"—William Burke and William Hare—decided to streamline their business, cutting out the intermediate stages of natural death and subsequent funeral (hence the English slang term "burking" for violent murder). When the scandal was uncovered and Knox was revealed as their principal (although supposedly unknowing) customer, he was immediately immortalized in schoolyard skipping rhymes, but his career lay in ruins.

Of necessity, the anatomist turned to his pen for support. Knox was a great enthusiast for morphology, but not for Cuvier's approach. He disdained final causes, referring contemptuously to the Bridgewater Treatises as the Bilgewater Treatises, and held forth against William Paley's vision of God as analogous to a watchmaker. Writing on the somewhat esoteric subject of "the structure of the stomach of the Peruvian llama," Knox warned that "it is not with animal machines as with a watch or other piece of human mechanism, wherein the purpose of its creation is expressly known and understood." He complained that the "animal machine abounds with structures, the reason for whose pressure he [the morphologist] cannot guess at, neither can he calculate what might be the result of their absence or destruction." Hence, "the attempts at particularizing the particular design connected with separate individual organs, seem to me hitherto to present a series of the most lamentable failures in human reasoning."[1]

The greatest influence on Knox's thinking about anatomy came from Germany, in the form of *Naturphilosophie* (also known as transcendentalism). Aristotle himself had noted that not all aspects of animal life seem readily explicable in terms of final cause—in particular, there are seemingly useless similarities (isomorphisms) between organisms of different species. Famously, he saw that soft-bodied animals (cephalopods) and hard-shelled animals (testaceous mollusks) are similar, if you ignore the shell on the latter. Many people had picked up on the isomorphic features of life—especially the functionally irrelevant similarities among the forelimbs of vertebrates—including early evolutionists like Erasmus Darwin, who suggested that they point to common ancestry. But much more than others, nineteenth-century German thinkers put morphological similarities at the center of their theory of nature. Isomorphisms (which later were called homologies) became an essential prop in a dynamic organic philosophy which saw all of nature as an interrelated whole.[2] Knox was highly influenced by this movement on the continent, and he was not alone.

Richard Owen, somewhat misleadingly known as the British

Cuvier, was a schoolmate of Whewell. By the 1830s, after receiving his medical training, he moved to London, where he rapidly assumed first rank among British anatomists.[3] Along with his passion for pure morphology, Owen was intensely engaged with embryology—especially comparative study of the developing forms of different species. The *Naturphilosophen* had long championed embryology because of the analogies it offered between the development of the individual and the development of the species. The natural world provided many counterexamples, however; and by the time that Owen was working, a somewhat more sophisticated position was on offer. This better position was the brainchild of the Estonian morphologist Karl Ernst von Baer.[4]

Like Cuvier, von Baer divided animals into four basic groups, which he subdivided according to similarities in embryological development. In his view, a primitive adult organism was not itself necessarily represented in the embryological development of a more sophisticated organism, but they might share common embryological forms. Since a primitive organism might not develop beyond its embryological state, von Baer speculated that the primitive adult form is part of the early growth of the complex organism. In Owen's own words: "Thus every animal in the course of its development typifies or represents some of the permanent forms of animals inferior to itself; but it does not represent all the inferior forms, nor acquire any of the forms which it transitorily represents."[5]

Tying this in with the history of life, Owen followed von Baer in seeing a movement from the general to the specialized. "The horse is the swifter by reason of reduction of its toes to the condition of the single-hoofed foot; and the antelope, in like manner, gains in speed by the coalescence of two of its originally distinct bones into one firm cannon-bone."[6] Trying to acknowledge both von Baer's idealism and Cuvier's teleology (that is, his search for final causes), Owen argued that in making an organism—for example, a vertebrate—there are two, somewhat conflicting forces at work. First, a "polarizing force" lays down the ground plan (or archetype, as Owen named it) of the basic organism, through a process he called "vegetative or irrelative repetition." In the vertebrate, for example, the archetype consists of a backbone made of components repeated over and over. At times, Owen referred to this archetype as something akin to a Platonic form, in that it gave the general pattern of the individual organism. Second, an "adaptive force" twists and transforms this archetype into specialized contrivances with direct functional ends—the beloved "design" of the natural theologians. The greater the effect of this second force, the further the organism moves from the original archetype.

Owen saw humans as the end point of this transforming process. At mid-century, he concluded a famous public lecture on limbs by making very clear his temporal ordering of the vertebrate world: "We learn from the past history of our globe that [nature] has advanced with slow and stately steps, guided by the archetypal light, amidst the wreck of worlds, from the first embodiment of the Vertebrate idea under its own Ichthyic vestment, until it became arrayed in the glorious garb of the Human form." Was this the thinking of an evolutionist? In the notebooks of one of Owen's close friends of the late 1830s, there are hints this might be so. "Mr. Owen suggested to me," Charles Darwin wrote, "that the production of monsters (which Hunter says owe their origin to a very early stage) which follow certain laws according to species present an analogy to the production of species." But these were private musings. Publicly, Owen drew no implications about actual connections through time.[7]

In this, he was much more discreet than some of his contemporaries. By the 1840s the idea of social progress had again taken off, and biological progress could not be far behind. Out of favor on both sides of the Channel after the French Revolution, progressivism began to revive first in France, with Claude Henri Saint-Simon's attempts to find and justify the laws of progress. It continued with his famous student Auguste Comte, who proposed a three-stage law: the theological, the metaphysical, and the positive or scientific. Around 1400 societies had progressed from the theological to the metaphysical, according to this interpretation of history, and now, in the nineteenth century—thanks in no small measure to Comte himself—social progress was moving from the metaphysical to the positive. There are obvious echoes here of the three-part temporal division of Joachim of Flora, and indeed the followers of Saint-Simon were ardent devotees of the medieval thinker.

Comte's toning down of the theological was very important as ideas like these found their way to England. There were folk in that land who had little or no formal religious belief and were primed to be enthusiastic recipients of positivism. The Scottish sage Thomas Carlyle and the empiricist English philosopher John Stuart Mill were two notable examples.[8] By the 1830s the country was ready again for a progressivist metaphysics, for by now the industrial revolution had got a second wind. Put on hold by the long conflict at the beginning of the century, thoughts of an ever-improving future were once again permissible. Especially important to this optimism was the coming of the railway, not only for the work that it provided in itself—squads of Irish navvies crisscrossing the country, digging tunnels, throwing up embankments, and terrifying the local populace on Saturday nights when they spent their pay—but for the rapid means of communication that it now opened to people of virtually every class. Cheap day-returns helped to make the Great Exhibition of 1851 a smashing success.

British deism had proven to be a hardy plant during the turn-of-the-century drought. One of its spectacular blooms was *Principles of Geology* written between 1830 and 1833 by the Scottish-born lawyer-turned-scientist Charles Lyell. Opposing the Providential intervention touted by the catastrophists, he argued that everything—mountains, lakes, rivers, plains, seas—can be explained in terms of unbroken laws acting over vast periods of time. But Lyell shared the catastrophists' concern that humans be seen as special. Here, uniquely, God had intervened with a miracle, he believed. Biological progressivism, if endorsed, would have forced Lyell toward evolutionism and that, in turn, would have led to the inevitable conclusion that humans too are a natural product of unbroken law.

To stop this fox, Lyell's "uniformitarian" theory (as Whewell labeled it) argued that organisms come and organisms go, without temporal direction or meaning. By way of proof, he gave a full exposition of Lamarck's false (as he saw it) theory of evolution, in an attempt to refute it. The effect was somewhat counterproductive. Those who read Lyell became, by default, acquainted with evolutionary ideas. More than this, Lyell had misunderstood Lamarck's evolutionism as offering an explicit explanation for the progressivism evident in the fossil record (which in fact Lamarck did not do). By going into this controversy in depth, Lyell again, inadvertently, helped his readers learn of an impressive fossil record that offered a powerful argument—certainly, the most psychologically convincing of arguments—for evolution.

Vestiges

The pump was primed. But if the idea of evolution was to gush forth, it needed a champion: a person able to write well and with enthusiasm for the subject, using analogies and metaphors and examples that would appeal to the general public, a person who was willing to push through to the end and see that the work was published and promoted. Such a person appeared in 1844 in the form of the Scottish publisher Robert Chambers. A very successful businessman—he and his brother built a media empire based on successfully gauging popular taste—Chambers produced a work that shocked, delighted, and left no one unmoved.[9] Written anonymously and destined to have major sales and to go through many editions, *The Vestiges of the Natural History of Creation* was the Big Mac of popular science—very tasty, very filling, very accessible, and (in the opinion of the authorities) of very dubious value to one's health. *Vestiges* was the archetype of pseudoscience. It was also the reigning paradigm of evolutionism, as opposed to mere evolution, at the middle of the century.

Chambers had always been interested in science, and he started out with the intention of writing a book on phrenology—the interpretation of bumps on the head as a guide to character. Somewhere along the line he switched to evolution, but without losing his gift for pseudoscience. Phrenology in Edinburgh around 1840 was a subject with strong ideological undertones.[10] Its enthusiasts pushed it as a modern, progressive way to think about humankind—a materialistic, somewhat deterministic way of thinking about personality and intelligence, as against the reactionary views of religious factions, which were committed to seeing humans as unconnected to their physical nature and as having autonomy and free will. Phrenology appealed to the person wanting to sweep away the old modes of acting and thinking and replace them with modern science, which, allied with proper training and education, would make for a better and happier society all around. Evolution, in Chambers's hands, was an idea of the same ilk, for the facts were secondary to the primacy of the ideology. Evolution, like phrenology, was a vehicle for pushing doctrines of progress.

Vestiges offered not just a story of biological evolution but a historical narrative of the entire universe. Chambers started with the evolution of the planets and then moved to the arrival of life through

spontaneous generation. As evidence for the latter, he offered the similarities between plants and the condensation patterns ("frost ferns")
on cold windows in winter. With that out of the way, Chambers
plunged into the fossil record.

Following Lyell's misreading of Lamarck but coming to the opposite conclusion, Chambers explicitly cited the fossil record as offering
support for evolution—from primitive forms up to humankind. "In
pursuing the progress of the development of both plants and animals
upon the globe, we have seen an advance in both cases, along the line
leading to the higher forms of organization." Plants go from seaweeds
up to the flowering plants of the land. Animals go from primitive
forms to fish and then up to reptiles. "From reptiles we advance to
birds, and then to mammalia, which are commenced by marsupialia,
acknowledgedly low forms in their class. That there is thus a progress
of some kind, the most superficial glance at the geological history is
sufficient to convince us." With a few German-inspired speculations
about causes thrown in, the picture was now complete. "The inorganic has one final comprehensive law, GRAVITATION. The organic, the other great department of mundane things, rests in like
manner on one law, and that is,—DEVELOPMENT. Nor may these
be after all twain, but only branches of one still more comprehensive
law, the expression of that unity which man's wit can scarcely separate
from Deity itself."[11]

"From Deity itself." Chambers, like many successful Scots, had
moved from the Presbyterian Church to the Episcopalian, but in truth
he never had much time for churchmen of any stripe. They represented the segment of society that he and his fellows were combating
and conquering. Demonstrating this bias—but also showing in the
process that he was no professional scientist—Chambers barely acknowledged function and ends, since they were, in his mind, the contrivances of the natural theologians. Lamarckian processes could do

whatever was necessary to produce the special features of organisms. His indifference to natural theology did not turn Chambers into an atheist or even an agnostic. He was, rather, a deist for whom the creation of organisms by lawful means was proof of God's existence rather than a refutation.

But more powerful than Chambers's unadorned deism was his commitment to the cultural notion of progress. It was the very root of his sense of self-worth and achievement. Chambers lived and breathed progress, and he explicitly saw evolution as a reflection of that ideal.

> The question whether the human race will ever advance far beyond its present position in intellect and morals, is one which has engaged much attention. Judging from the past, we cannot reasonably doubt that great advances are yet to be made; but if the principle of development be admitted, these are certain, whatever may be the space of time required for their realization. A progression resembling development may be traced in human nature, both in the individual and in large groups of men . . . Now all of this is in conformity with what we have seen of the progress of organic creation. It seems but the minute hand of a watch, of which the hour hand is the transition from species to species. Knowing what we do of that latter transition, the possibility of a decided and general retrogression of the highest species towards a meaner type is scarce admissible, but a forward movement seems anything but unlikely.[12]

Those who endorsed Chambers's social philosophy—his ideology of change—generally accepted his thinking about evolution, moving along with him to a full-blown evolutionism. Even religious thinkers like the Reverend Baden Powell (father of the future founder of the

Boy Scouts)—people whose religious beliefs inclined them toward a God who works through unbroken law, a God for whom human-driven progress was not antithetical to his world scheme, that is, the God of the postmillennialist—responded positively to *Vestiges.* Likewise, the poet Alfred Tennyson. *In Memoriam*—a series of poems dedicated to his too-young-dead friend Arthur Hallam, over which he labored for twenty years—is infused with the science of the day.[13] A student of Whewell, Tennyson knew his geology and found the uniformitarian message of Lyell's *Principles* deeply depressing. All that energy, all that conflict, all of that effort adding up to nothing, going nowhere.

> Are God and Nature then at strife,
>   That Nature lends such evil dreams?
>   So careful of the type she seems,
> So careless of the single life . . .
>
> So careful of the type? but no.
>   From scaped cliff and quarried stone
>   She cries, "A thousand types are gone:
> I care for nothing, all shall go."

Given Nature "red in tooth and claw"—this famous phrase has its source here—nothing seems to make any sense. There are just endless Lyellian cycles.

Tennyson began *In Memoriam* in 1833, but after six years of work he felt unable to move forward. Then, in the mid-1840s, he read *Vestiges.* Chambers's unambiguously progressionist organic evolution inspired the poet, who picked up his pen and finished his work. Tennyson argued in the final lines that perhaps we have meaning after all, despite a Lyellian uniformitarianism: life is progressing upward, and perhaps will go on beyond the human form that we inhabit at present.

Could it not be that Hallam represented some anticipation of a more-developed life to come, cut short in its prime? Therefore, hope exists for us all, and Hallam's life has meaning.

> A soul shall strike from out the vast
> And strike his being into bounds,
>
> And moved thro' life of lower phase,
>   Result in man, be born and think,
>   And act and love, a closer link
> Betwixt us and the crowning race . . .
>
> Whereof the man, that with me trod
>   This planet, was a noble type
>   Appearing ere the times were ripe,
> That friend of mine who lives in God.

Not everyone felt this way about Chambers's vision. The Providentialists hated *Vestiges*. For Adam Sedgwick, the book was the work of the Antichrist, and its popularity—especially with the ladies—only made matters worse. Having at first speculated that a work so bad must have been written by a woman, he then pulled back, declaring that no woman could be so debauched as to write such an appalling work: "The ascent up the hill of science is rugged and thorny, and ill suited for the drapery of the petticoat." The Scottish scientist David Brewster likewise moaned about this aspect of the work: "It would auger ill for the rising generation, if the mothers of England were infected with the errors of Phrenology: it would auger worse were they tainted with Materialism." The problem, Brewster gloomily concluded, was with the slackness of today's classrooms. "Prophetic of infidel times, and indicating the unsoundness of our general education, 'The Vestiges . . . ' has started into public favour with a fair chance of poisoning the fountains of science, and of sapping the foundations of religion."[14]

Whewell particularly took umbrage with the fact that *Vestiges* denied—or ignored, at least—final cause. Refusing to accord *Vestiges* the compliment of a published refutation, he gathered together pieces from his own Bridgewater Treatise and his two works about science, the *History* and the *Philosophy of the Inductive Sciences,* and with a stirring introduction (the first version of which did not mention *Vestiges* by name) published them under the revealing title *The Indications of the Creator.* The book brimmed with assertions about the all-important significance of final cause.

But the damage was done. Even if *Vestiges* was wrong, the idea of evolution, like the idea of progress, was now a familiar fixture on the intellectual and popular landscape. By the 1850s in Britain, whether the establishment scientists wanted it that way or not, evolutionary ideas were becoming known and less threatening. They could even be inspiring. And that was no small thing. They were hardly in any sense respectable, judged as professional science. But they were no longer so tense-making.

On the other side of the ledger, no matter how good a new idea may be, people will not accept it so long as the old ideas function strongly and positively. Many religious people in Britain still opposed the idea of evolution. But how strong and firm was this opposition, around the time that Owen put forward his idealist theory and Chambers wrote and published *Vestiges,* and before a new and much stronger evolutionary contender burst on the scene? To answer that question, we will look first at the evangelicals or "low church," then at the other end of the spectrum, Catholic ideas and practices ("high church"), and finish with those people, particularly scientists, in the middle, who were trying hard to reconcile faith and reason, religion and science ("broad church").[15] For now, we will focus on Britain, since America by mid-century was too separate, too different, and too important to be rolled into a discussion of the old country.

Although generalizations are dangerous, one could say that in Britain around this time most dignified Christians had pulled back in embarrassment from detailed apocalyptic speculations. To a great degree, this was less a matter of theology than of class. In the lower levels of society, an ongoing stream of prophets and prophetesses (most notoriously, at the beginning of the nineteenth century, the farmer's daughter Johanna Southcott) claimed insights into the nature of reality and forecasted the end of time.[16] Premillennialism—the belief in Christ's second coming at the beginning of the millennium—rode high among the socially down-trodden and will have important implications for our story. In the middle and upper class as well, premillennialists were not totally unknown. The only British prime minister ever to be assassinated, Spencer Perceval (killed in 1812), was a premillennialist. So also was Anthony Ashley, the Seventh Earl of Shaftesbury, the great reformer of factory conditions. But generally speaking, respectable Britons (especially those who were going to take an interest in the relation between science and religion) simply did not indulge in this sort of thing. Speculations for or against the Whore of Babylon would not have gone down well at the Royal Society of London.

Evangelicals and Science

Evangelicals had a bad reputation in nineteenth-century England. Novelists cruelly caricatured them as kill-joy hypocrites. In Charles Dickens's *Pickwick Papers*, the Reverend Mr. Stiggins is always looking for free food and drink (especially the latter); in Anthony Trollope's *Barchester Towers*, the unctuous Obediah Slope, the bishop's secretary, is plotting to run the diocese and marry the Widow Bold. But particularly north of the border, in Scotland, where education was traditionally valued highly, the most enthusiastic evangelicals were often among those who pursued science most vigorously, thinking that thereby they were revealing the glory and greatness of the Creator. Thomas Chalmers, the leader of a group who left the Church of Scotland in 1843 because they wanted freedom to choose their own ministers, lectured frequently on science and was respected for his understanding of the serious issues of the day.

One can find people in England who pushed for a fairly stern reading of the Holy Book. But in the early part of the century even most evangelicals recognized that some accommodation had to be made to modern science. At the end of the sixteenth century, based on the genealogies given in the Bible, Archbishop Ussher had worked out that the earth must be about 6,000 years old. Clearly, in the light of subsequent geological discoveries, this time period had to be extended, drastically. One could do this either by supposing that after the initial creation there were many years of unrecorded history, before God got back into the creative acts of the early chapters of Genesis. Or one could simply extend the six days indefinitely, by reading them as six long periods of creative effort.

Probably the most successful and widely read of the "ages theorists" around mid-century was Hugh Miller, a Scot who began life as a stonemason (hence the interest in geology). A leading evangelical who followed Chalmers out of the Church of Scotland, he became highly respected by professional geologists and also a very popular writer, especially on scientific matters. Miller went to some lengths to show that the geological record, including the fossil record, fit with the kind of account presented in Genesis, so long as one interpreted the story in terms of periods rather than literal days. He was prepared to back his case with a theological argument: "God the Creator, who wrought during six periods, rested during the seventh period; and as we have no evidence whatsoever that he recommenced his work of creation,—as, on the contrary, man seems to be the last formed of creatures,—God may be resting still. The presumption is strong that his Sabbath is an extended period, not a natural day, and that the work of Redemption is his Sabbath-day's work."[17]

Miller was no friend of evolution. The Bible spoke against it, he said, natural theology spoke against it, and science spoke against it. But it was very important for Miller, and for those who devoured his books, that the case against evolution be made on scientific grounds. Much effort was made to demonstrate that the fossil record absolutely negates transmutationary hypotheses. There are far too many gaps between forms to suppose gradual change. Moreover, when new forms do appear in the record, they tend to be highly complex and sophisticated. Focusing on the remains of a fossil fish *(Asterolepis)* that he himself had discovered, Miller drew a happy conclusion: "Up to a certain point in the geologic scale we find that [fish of this kind] *are not;* and when they at length make their appearance upon the stage, they enter large in their stature and high in their organization."[18] This evidence showed that God worked by direct, out-of-the-order-of-nature intervention.

Yet to make his case, not only was Miller appealing to science—and therefore at least laying open the possibility that new discoveries might demand a change of mind—but often he was appealing to the very science that gave evolutionists cause for celebration. Although he would not have given a progress-based interpretation, Miller was as ardent a biological progressionist as Chambers, seeing God's creative work as reaching its climax with our own species. Like Chambers also, and before him the *Naturphilosophen,* Miller saw parallels between embryological development and the history of life, and he thought this pattern was highly significant. Later, he even went so far as to allow that just before the reptiles arrived on the scene, the fish were at their most reptilian!

Hugh Miller went to his grave thinking evolution a vile and untenable hypothesis. Looking back over a span of a century and a half, one can be forgiven for thinking that evangelicals like him were doing more to reverse their own position than any force from without.

John Henry Newman

Lead, kindly Light, amid th' encircling gloom,
Lead Thou me on!
The night is dark, and I am far from home;
Lead Thou me on!
Keep Thou my feet; I do not ask to see
The distant scene; one step enough for me.

This opening stanza of the Victorians' favorite hymn was written in 1833 by the young John Henry Newman, as he was making his way home to England from an extended stay in Italy. At one level, the words were literally autobiographical, as the writer yearned for his homeland and wanted desperately to return. At another level, the words were metaphorically autobiographical, as Newman alluded to the religious journey he was then making, which was taking him from an evangelical childhood, up through the levels of Anglicanism as he became more and more catholic and high church, until finally in 1845 he made the leap over to Rome, eventually at the end of his long life becoming a cardinal in the Roman Catholic Church. Although Newman always had an interest in science—as an undergraduate at Oxford he had attended (and enjoyed) Buckland's lectures on geology—his journey was in no way driven by science, for or against. Yet the thinking of this great theologian had implications for anyone wrestling with ideas of evolution and change. It would not lend simply positive support to science, but it would not be simply negative either.

Newman went to Oxford in pursuit of the middle way of the established church, between Puritanism and Catholicism. But that route did not provide a satisfactory and stable answer to his quest.[19] Newman and his friends went on to form the Oxford Movement, which published a series of pamphlets intended to reform the Anglican Church and remind its members of its roots and obligations in the Catholic tradition. The most notorious of these tracts was the final one, *Tract XC*, where, to the astonished disgust of many of his Anglican co-religionists, Newman essentially rejected the Reformation, arguing that nothing in Anglicanism is contrary to the stand of the Roman Church. But even before this, Newman had been sawing away at the supports of the Reformation. To those who considered the Bible the foundation of their religion, Newman took on the sacred text in *Tract LXXXV.* With a zest worthy of today's secular humanists, he demonstrated that the Bible is a frail reed indeed on which to base one's faith. There are two stories of creation in Genesis, for a start. In the one, Adam and Eve are created together. In the other, Eve is made from Adam's rib, almost as an afterthought. In Deuteronomy, there are two accounts of Moses fasting on the Mount, in Exodus only one. God told Balaam to go with the visitors, and then God was angry precisely because he went. Manasseh used enchantments and summoned spirits and wizards, but then he repented and was forgiven; why then does God keep punishing people later because of Manasseh's sins? The Gospels were not much better. Was it four thousand people fed with seven loaves or five thousand with five loaves? Did Simon of Cyrene carry the Cross, or did Jesus carry it? And how about Judas? Did he hang himself, as Matthew tells us, or did he fall headlong with his bowels gushing out, as we learn in the Acts?

If revealed religion based on literal readings of the Bible was weak, what of natural theology? Newman was little more enthused by this.

In 1870 (twenty-five years after he converted to Catholicism), in correspondence about his seminal philosophical work, *A Grammar of Assent*, Newman wrote: "I have not insisted on the argument from *design*, because I am writing for the 19th century, by which, as represented by its philosophers, design is not admitted as proved. And to tell the truth, though I should not wish to preach on the subject, for 40 years I have been unable to see the logical force of the argument myself. I believe in design because I believe in God; not in a God because I see design."[20] He continued: "Design teaches me power, skill and goodness—not sanctity, not mercy, not a future judgment, which three are of the essence of religion." Newman was simply not in the business of using science to support his religion. And by the same token, in no way could science be a threat to his religion.

So where did Newman find the grounding of his faith? In tradition! He argued that true doctrine is something revealed gradually through the ages, starting with Christ but developing slowly and sequentially as followers pondered on these things and truths were uncovered. Original sin, the Trinity, atonement, eternal salvation—these are all things that are not to be found in the Gospels but in the thinking and writings of Saint Paul and then of the Church fathers, notably Augustine. Christianity was something that grew organically, one might say.

Did this make Newman a kind of simple-minded cultural evolutionist?[21] Although at some level Newman's Christianity progressed from the simple to the complex, his theological vision was deeply Providential. God guided the uncovering. And Newman would never think of the very old as being finally discarded as the new pushed it aside. As one might say—and as Newman did say, in a work published in 1845, marking his move from Canterbury to Rome—Christian doctrine should be thought of as a type (a pattern or blueprint), and the change that occurs is in some way a fleshing out and elaboration of that type. "That the essential idea or type which a philosophical or political system represents must continue under all its development, and that its loss is tantamount to the corruption of the system, will scarcely be denied."[22]

This kind of thinking—akin to the morphological thinking of Richard Owen that made much of archetypes—filled out Newman's ideas on doctrinal development. It led up to the present, with a world more fully and better understood than it was before. Not that we have progress. Newman always hated this idea, thinking that the world of the Victorians was sinful, cruel, godless. Temperamentally, Newman had little time for millennial speculations of any kind. But he knew what he disliked more than what he liked. He loathed the new vogue for postmillennialism—that secularized version popular among philosophers and social theorists—and typically preferred the more intervening-God-up-front superstitious premillennialism to nothing at all. Better "he, who from love of Christ and want of science, thinks some strange sight in the sky, comet or meteor, to be the sign of His coming, than the man, who, from mere knowledge and from lack of love, laughs at the mistake."[23] Hence, Newman could never have endorsed evolutionism.

Although Newman could no more have bought into a blind materialistic form of creation than could the crude literalists we will encounter later, one who thought as he did could go some considerable way toward accepting a form of evolution—at least the fact of evolution, that is. Neither the Bible nor final cause would stand in the way, and a significant kind of developmentalism had already been endorsed by him.

The Anglican Scientists

Finally, what of those scientists trying desperately to have their faith and their reason also? They found their own antievolutionary thinking coming apart in their hands, with no scientific counterproposal to offer. The Oxbridge scientists particularly had been much involved in running professional scientific organizations such as the Geological Society of London and the "British Ass" (British Association for the Advancement of Science), which held an annual meeting for one and all in different provincial towns. Whewell, Sedgwick, and others, especially the astronomer John F. W. Herschel, had also been hard at work trying to define the nature and limits of professional science. While there were differences in approach, all agreed that good science is like Newtonian mechanics, a lawbound axiom-based system, incorporating what Newton somewhat mysteriously had labeled *verae causae*, true causes.

Whewell, for his part, took something of a rationalist approach, arguing that direct sensory evidence of such causes was unnecessary, that they could be inferred indirectly from conclusions, especially when such conclusions are what he called "consilient"—consistent across different branches of knowledge and unified by this shared cause. Herschel took something of an empiricist approach, arguing that one must either see such causes or have analogical evidence of their existence and workings. But the differences were less than the agreements—law and cause. And now Whewell was loftily claiming that the end-directed nature of organisms makes impossible any natural explanation—any explanation in terms of *verae causae* within a law network. No scientific answer will satisfy. "Indefinite divergence from the original type is not possible; and the extreme limit of possible variation may usually be reached in a short period of time: in short, *species have a real existence in nature*, and a transmutation from one to another does not exist . . . The mystery of creation is not within the range of her legitimate territory; she says nothing, but she points upwards."[24]

A neat way out, but by the mid-nineteenth century not really good enough. And this was only one of the things troubling these men, who were trying to stay within a traditional form of Protestant Christianity and also go the full way with modern science. Morphology was a real problem. Sedgwick and Whewell seized gratefully upon Owen's archetype theory. They would have preferred to see Cuvierian influences as dominating all, but if this could not be so, then Owen provided a respectable alternative. Yet, the simple fact was that Owen's theory put one on a slippery slope to "transmutation," that is, to evolution. Why not suppose that the archetype was the ancestor and all else was molded through development from it? Sedgwick spoke frankly about the dangers of any transcendentalist morphological speculations: "I detest them, because I think them untrue. They shut out all argument from *design* and all notion of a Creative Providence."[25]

Whewell, the polymath, always one step ahead in sensing trouble, saw even greater worries, as he started to obsess about an old but perennial cosmological problem, the so-called plurality of worlds: is there intelligent life elsewhere in the universe? If so, did this mean that we humans are no longer uniquely favored of God? Was the message of Christian revelation being diluted? If not, then for what purpose did God create all of those endlessly circulating lifeless globes? Does this not strike at the heart of the design argument, showing that much in the world is without purpose, and hence we should not be so confident about final cause down here?

Whewell had a solution, namely that we are unique—Christian revelation is preserved—and that God put the other planets and suns and stars in place for us to discover his magnificent creative abilities. "If God have placed upon the earth a creature who can so far sympathize with Him, if we may venture upon the expression . . . then, so far as intellect alone (and we are here speaking of intellect alone) can make Man a worthy object of all the vast magnificence of Creative Power, we can hardly shrink from believing that he is so." No one was really convinced by this. One wag suggested that Whewell's true motivation was to prove that "through the whole of infinity / there is none so great as the Master of Trinity."

Ambition aside, by stressing the ubiquity of law, Whewell embraced the very explanation he wanted to deny in the case of organic origins. "The planets and the stars are the lumps which have flown from the potter's wheel of the Great Worker;– the shred-coils which in the working, sprang from His mighty lathe;– the sparks which darted from His awful anvil when the solar system lay incandescent thereon;– the curls of vapour which arose from the great cauldron of creation when its elements were separated." Even those who shared Whewell's antipathy to evolution thought this pretty weak.[26]

The fact is that in Britain, by the mid-nineteenth century, there simply was no united wall of opposition to the kinds of thinking that evolution represented. Even the deeply religious—perhaps especially the deeply religious—were themselves pointing to ways in which a transformist hypothesis might secure a beachhead in the respectable scientific world. But before anything of this nature could happen, a fresh approach was needed, one that offered a natural explanation of organic origins, yet measured up to the standards of good science— an approach that would speak to adaptation and final cause and at the same time be sensitive to homology. Such an approach was about to be made by Charles Darwin.

4

Charles Darwin

London. Michaelmas term lately over, and the Lord Chancellor sitting in Lincoln's Inn Hall. Implacable November weather. As much mud in the streets as if the waters had but newly retired from the face of the earth, and it would not be wonderful to meet a Megalosaurus, forty feet long or so, waddling like an elephantine lizard up Holborn Hill. Smoke lowering down from chimney-pots, making a soft black drizzle, with flakes of soot in it as big as full-grown snowflakes—gone into mourning, one might imagine, for the death of the sun. Dogs, undistinguishable in mire. Horses, scarcely better; splashed to their very blinkers. Foot passengers, jostling one another's umbrellas in a general infection of ill temper, and losing their foot-hold at street-corners, where tens of thousands of other foot passengers have been slipping and sliding since the day broke (if this day ever broke), adding new deposits to the crust upon crust of mud, sticking at those points tenaciously to the pavement, and accumulating at compound interest.

*D*inosaurs fascinated and terrified the Victorians as much as they do us. Richard Owen introduced the idea in the early 1840s, when he described this newly found type of long-extinct reptile and gave it a class and name of its own. The new group was an immediate success. Just as we today have dinosaur themes and pictures in our popular culture (even Dinoghetti, canned pasta in tomato sauce), so the Victorians had their dinosaur themes and pictures. On one memorable occasion, earnest scientific dignitaries gathered in a hollow iguanodon to eat dinner. The Great Exhibition of 1851 featured rather fanciful plaster models, including a rhinoceros-like megalosaurus which to this day can still be seen in the park at the Crystal Palace in South London. Charles Dickens was ever sensitive to popular culture—he had read Chambers's *Vestiges* and become an ardent evolutionist—and in 1852 he drew naturally on the dinosaur image for the opening of his great novel *Bleak House,* where he painted a picture of sooty, muddy London, with its dreadful winter, as a throw-back to some prehistoric clime and time.

The Victorians were ready for a breakthrough. Thanks to Chambers, Tennyson, and others, evolution was in the air. And thanks to Owen, Dickens, and others, dinosaurs and other long-extinct brutes were part of popular culture. Progress was an ideology whose time had finally come. As "Dover Beach" makes clear, faith was retreating, and somebody had to make sense of all of this. Charles Darwin was ready to take up the task.

Born in the town of Shrewsbury in the British Midlands on February 12, 1809 (the same day as Abraham Lincoln's birth), Charles Darwin was the fifth child of Dr. Robert Darwin, himself the eldest son of Erasmus Darwin. A very successful physician, Robert Darwin was also a money man, serving as an intermediary between aristocrats who needed to mortgage their land holdings for cash and industrialists who needed safe harbors and profitable investments for their earnings. Charles Darwin's mother, who died before he was ten, was the daughter of Erasmus Darwin's fellow "lunatick" Josiah Wedgwood, the very successful leader of Britain's pottery industry. From birth, young Darwin belonged to the rich upper-middle classes, and his financial comforts were further guaranteed when at the age of thirty he married his first cousin, Emma Wedgwood (daughter of Josiah Wedgwood's son, also called Josiah).

The young couple moved to the Kentish countryside, in the village of Downe, where they chose a home called, somewhat confusingly,

Down House (without the final "e"). There, they raised their children, seven of ten surviving to adulthood. For some unknown reason, about the time of his marriage Darwin fell sick and for the rest of his life suffered constant headaches, stomach pains, and other strange symptoms. The Darwins—deeply committed to each other and to family—became reluctant to socialize outside their home, although through the years many people came to visit them. Charles Darwin died of a heart attack at age 73, on April 19, 1882. By popular demand of the scientific community and with the willing agreement of the Anglican Church, he was buried next to Isaac Newton in Westminster Abbey, a testament to his considerable contributions to science.[1]

Darwin's status and the money are worth stressing right from the beginning. Charles Darwin was never in financial need, and so could devote his attentions to a life of science without the distraction of gainful employment or the fear of poverty from making a false move. Yet when he wrote his many books, including the *Origin of Species*, he would always have, at the back of his mind, the views of his patrons, particularly his father and uncle. Darwin was a great scientist, a very great scientist, whose fertile brain produced one hypothesis after another, but one should not expect him to reject his upbringing. He might transform it or give it new meaning, but it would be there in the background nevertheless.

If background was so important, from what belief system did Charles Darwin emerge? Although Robert Darwin's religious faith was minimal to the point of nonbeing, he raised young Charles as an Anglican and sent him off to the University of Edinburgh, where he was expected to follow in his father's footsteps and become a doctor. After two years of study during which apparently he showed no great motivation or talent for medicine, Charles moved to the University of Cambridge, intending now to become an Anglican clergyman. This was the ideal post for a young man with money but without a great deal of drive or obvious ability.

Darwin did not take a science degree because in those days there were no science degrees, but he did attend the botany lectures of John Henslow and obviously impressed his seniors (including Adam Sedgwick and William Whewell) with his passion for natural history and his determination to make some kind of mark. Upon graduation in 1831, young Darwin was offered the position of captain's companion on board *HMS Beagle*, just then embarking on what was to prove a five-year stint of charting the coasts of South America and eventually going all the way around the world. He accepted the offer and rapidly evolved into the ship's naturalist, making massive collections of birds, plants, fossils, and more, all of which were packed up and shipped back home. Darwin wrote up an account of his voyage that delighted not just his own family but countless of his fellow Victorians. Even if he had never become a scientist, Darwin would have been loved and respected as one of the great travel writers of the age.

When he set off from England, young Charles was a committed Christian, still intended for a life in the Church upon his return. His shipmates even joked about his rather literal approach to the Bible. His views started to change during the course of the journey, however, and by the time he returned in 1836 he seemed to have become more of a deist. The move to deism was no big shock. Many people in the Darwin family were deists, and the Wedgwoods were of the Unitarian faith, which had at its center the denial of the Trinity. But the spur to Darwin's shift from natural theology to deism was probably Charles Lyell's *Principles of Geology*. Darwin took the first volume (published in 1830) with him when he left, and as the other two volumes appeared they were sent out to him by ship. Seduced by Lyell's vision and his easy style, Darwin became an ardent uniformitarian.

Darwin's first independent scientific research was done very much as an adjunct to Lyell's larger project. The older man was puzzled by coral reefs, and he suggested that they were located on the rims of extinct volcanoes. Darwin saw at once how improbable this hypothesis must be. Lyell was supposing that each and every volcano independently grew up from the sea bed to just such a height that it broke the surface of the sea, where the coral could grow and flourish. As part of his uniformitarian theory, in order to account for climatic fluctuations on earth, Lyell had suggested that the earth's surface is constantly rising and falling—when one area sags, another rises, rather like a water bed. Taking a leaf from Lyell's own book, Darwin suggested that the volcanoes had been of various heights but were on falling sea beds. Just as the rims were about to vanish beneath the surface, the coral could attach itself and grow, and then as the volcanoes fell further, the living coral simply built on dead forms to stay at sea level.[2]

More significant than the science as such was the fact that, as he was filling in a piece of the Lyellian picture, Darwin was buying into the methodology, which downplayed or dismissed miracles as unnecessary, given the operation of natural laws and immense periods of time. By minimizing miracles, Darwin delivered a heavy blow to his own Christian beliefs. When he was at Cambridge, the received authority on revealed theology had been William Paley's *Evidences of Christianity*—a book that based the whole rationality of the Christian drama on the existence of miracle. With no miracle, there was no Christian faith.

What of evolution? Darwin had been acquainted with evolutionary ideas from an early age. At some point he had read his grandfather's major work, *Zoonomia*, and when he was at Edinburgh he had mixed with at least one man favoring evolution (Robert Grant). Then there was Lyell's discussion of Lamarck. Yet (underlining the pattern that evolution is more an effect than a cause of loss of faith) Darwin seems not to have become an evolutionist on the *Beagle* voyage. But one thing that did fascinate him was biogeography, the distributions of organisms around the globe. When in 1835 he visited the Galapagos Archipelago, a group of Pacific islands on the equator, he was most excited to learn that the giant tortoises on the islands—some living within ten miles of each other—were of different kinds. He then discovered for himself that the same seemed true of the islands' small birds. This fact struck him as particularly important. On the South American continent, he had found birds that were quite similar, even though they ranged over vast distances and lived in very different climates.

None of this made sense—neither scientifically nor theologically. Darwin had also read Paley's *Natural Theology* at Cambridge and had been much impressed. Why would God have lavished so much care on the Galapagos, making the reptiles and birds there unique from island to island? When he returned to England and an ornithologist confirmed that the birds are indeed of different species, Darwin grasped the nettle of evolution. He was looking for a lawbound way to explain the variations, and evolution seemed the only answer. But what was the mechanism that caused these changes? For some eighteen months Darwin worked hard on the problem, until in the fall of 1838 he hit on his solution—the natural selection of the favored (or fittest) brought on by the struggle to survive and reproduce.

One might say that Darwin's shift from Christianity to deism was decisive in his becoming an evolutionist, although the influence of his grounding in natural theology cannot be discounted. But when he became not just an evolutionist but a Darwinian—that is, when he embraced natural selection—his Christian background reasserted itself. Most particularly, Darwin accepted—he had always accepted—the Christian claim that the organic world functions as if it were designed, and that these "designed" adaptations or contrivances are necessary for survival and reproduction. He was with Paley and Cuvier

and Sedgwick and all of the others on this. These conditions of existence (what Darwin tended to call "Conditions of Life") were primary for him. He knew about homology (which he called "Unity of
Type") and was not about to dismiss its importance, but it was an af-
ter-effect of evolution, not a central issue. In other words, for Darwin
any evolutionary mechanism had to explain adaptive complexity. Darwin did not now want to deny the argument for a designer, but the designer had to work at a distance, through and only through the law of
natural selection.

The first breakthrough on the route to natural selection was realiz-
ing that animal and plant breeders get their successes through selection—picking those individuals with characteristics the breeders want
and then breeding them, while killing off (and sometimes eating)
those individuals with traits that are less desirable. An industrial revolution demanded an agricultural revolution, and Darwin—coming
from rural Midlands—knew all about these sorts of things. His family kept pigeons and his Uncle Josh bred sheep, and everyone knew of
the great successes in agriculture, especially the development of tur-
nips and other root crops that enabled farmers to keep their livestock
alive through the winter.

These breeding practices were all about design. No less than the
ironmaster who fashioned the blast furnace that he wanted, the
breeder fashioned the animal or plant that he wanted, using selection.
A natural form of this selection would produce precisely the contrivances that the natural theologians found so vital. But one still had to
explain how natural selection could ever get from the desk and to the
field. What would force selection to work nonstop, continually mold-
ing plants and animals into new species? Human selection seemed always to have limits and a tendency to revert whenever there was a
pause in breeding. Why would the natural world not behave similarly?

At the end of September 1838 Darwin read Robert Malthus's

Essay on a Principle of Population, and the other piece of the puzzle suddenly fell into place. Malthus, another Anglican clergyman and a Providentialist, had been concerned to counter what he felt was false hope so many were harboring about progress and the possibility of arriving at utopia. At the same time, he wanted to answer a question that had puzzled many Christians: How is it part of God's design that we humans should make any effort at all? Why is it not part of human nature just to hang around doing nothing? And if it is part of human nature that we hang around and do nothing, why do we not do so?

Malthus's solution came in a famous formula: population numbers tend to increase geometrically (1,2,4,8, . . .), whereas food supplies increase at most only arithmetically (1,2,3,4, . . .). This means that demand eventually outstrips supply and that "struggles for existence" are inevitable, leading to the death of many individuals and species. Indefinite upward change is impossible because of this failure to survive and reproduce. In a later, more moderate version, Malthus allowed that "prudential restraint" in human reproduction and the rationing of crops might make the struggles less than inevitable, but neither he nor anyone else really thought that these steps were ever likely to be taken on a significant scale. After reading Malthus, Darwin realized that no such restraint and rationing was possible in the nonhuman world, and the struggle for existence that inevitably occurred was the engine powering a natural form of selection. Organisms that were less well-equipped for the struggle would not survive or reproduce, and the offspring of better-adapted organisms would dominate the next generation. Darwin thus turned Malthus's position on its head, from no change to permanent, never-ending change.

Darwin discovered natural selection in 1838. In 1842 he sat down and wrote a short sketch of his ideas, and then two years later he wrote a more detailed essay, which he showed to his botanist friend

Joseph Hooker. He did not publish either of these, although he left very careful instructions for the publication of the latter were he to die prematurely. Darwin was finally pushed into print when a young Welsh-born naturalist, Alfred Russel Wallace, sent him an essay in 1858, in which Wallace described his own independently discovered theory of evolution through natural selection—almost an exact replica of Darwin's own thinking.[3] At the insistence of Lyell and Hooker, Darwin set to work immediately to collect enough of his already-written material to establish priority. Lyell and Hooker presented both men's papers to the Linnaean Society, and the next year, in 1859, having written up his ideas in a book-length manuscript, Darwin published *On the Origin of Species by Means of Natural Selection, or the Preservation of Favoured Races in the Struggle for Life.*

The Origin of Species

The *Origin* was a carefully crafted book. Darwin's mentors—Henslow, Sedgwick, Whewell—had made sure that he knew the rules of professional science. They had urged him to read Herschel on the nature of science, and he knew that one must aim for explanations of the natural world based on a network of physical laws. That is what Darwin attempted to offer in the *Origin*, going first to the struggle for existence:

> A struggle for existence inevitably follows from the high rate at which all organic beings tend to increase. Every being, which during its natural lifetime produces several eggs or seeds, must suffer destruction during some period of its life, and during some season or occasional year, otherwise, on the principle of geometrical increase, its numbers would quickly become so inordinately great that no country could support the product. Hence, as more individuals are produced than can possibly sur-

vive, there must in every case be a struggle for existence, either one individual with another of the same species, or with the individuals of distinct species, or with the physical conditions of life. It is the doctrine of Malthus applied with manifold force to the whole animal and vegetable kingdoms; for in this case there can be no artificial increase of food, and no prudential restraint from marriage.

And then on to natural selection:

Let it be borne in mind in what an endless number of strange peculiarities our domestic productions, and, in a lesser degree, those under nature, vary; and how strong the hereditary tendency is. Under domestication, it may be truly said that the whole organization becomes in some degree plastic. Let it be borne in mind how infinitely complex and close-fitting are the mutual relations of all organic beings to each other and to their physical conditions of life. Can it, then, be thought improbable, seeing that variations useful to man have undoubtedly occurred, that other variations useful in some way to each being in the great and complex battle of life, should sometimes occur in the course of thousands of generations? If such do occur, can we doubt (remembering that many more individuals are born than can possibly survive) that individuals having any advantage, however slight, over others, would have the best chance of surviving and of procreating their kind? On the other hand we may feel sure that any variation in the least degree injurious would be rigidly destroyed. This preservation of favourable variations and the rejection of injurious variations, I call Natural Selection.[4]

But Darwin had to do more than this. He had to show that his theory was causal, and that natural selection was the true cause. Darwin was sensitive to the differences between the empiricist approach and the rationalist approach. The 1830s were a time of intense discussion in Britain about the relative merits of the particle theory of light and the conquering wave theory (known then as the undulatory theory). The trouble was that no one could see the particles or waves, and Herschel and Whewell were trying to show how their respective philosophies dealt better with the issues—Herschel trying to find physical analogies for waves, and Whewell insisting that one simply had to show that the wave hypothesis was the center of a unifying consilience. Analogously, in their debate over geology, Herschel opted for a Lyellian approach, whereas Whewell insisted that we have no need to see catastrophes directly so long as they explain the evidence at hand.

Being clever and much concerned to elevate his evolutionary theory from its pseudo status, Darwin tried to show how natural selection satisfies both Herschel's empiricist criteria and Whewell's rationalist criteria. He started by exploiting the analogy between artificial and natural selection—going from the one, the known and perceived, to the other, the unknown and inferred. The analogy is a commonplace for us today, but in Darwin's time it was quite bold. In fact, back then the comparison with breeding was the standard argument *against* evolution (the farmer does not turn a pig into a cow, it was said), so much so that when Wallace wrote up his own essay, he spent much time explaining why the two processes are not comparable—and why artificial selection should not be taken as a paradigm for nature.

Darwin argued—in true Lyellian spirit—that time, which can make mountains out of molehills, can mold any species on earth. Probably this analogy led Darwin also to a secondary mechanism, sexual selection, which comes not from a struggle for existence but from a struggle for mates. Breeders pick ever stronger and more vicious fighting cocks and ever more beautiful and tuneful birds, and so likewise does nature, because the ability to defeat competitors and attract suitors is critical for reproductive success in many species.

Having presented the empirical evidence for evolution by natural selection, Darwin next made the rationalist argument for consilience. He worked through the various branches of biology, showing that selection throws light on previously unexplained problems, and conversely that what is well known in these fields lends credence to evolution. He started with instinct, arguing that the intricate and complex relationships between social insects are best understood in terms of adaptive advantage, and that we have enough evidence of variation and struggle to infer the operation of selection. Next came geology and paleontology. In many respects these fields provided wonderful evidence for evolution, and Darwin knew this, but some explaining was necessary to account for the gaps in the record. Also, back then, little if anything was known about the evolution of life before the rather complex forms of the Cambrian burst into the record, as if created on the spot, and this too required some explanation. Next, Darwin moved to geographical distribution, or biogeography, which got loving treatment owing in large part to Darwin's personal experiences on board the *Beagle*. He was particularly keen to stress the way in which island life can be explained through the natural forces of immigration—wind and driftwood and the like—together with evolution after the founders have arrived. Moving to morphology, he showed how homology follows at once from the evolutionary picture.

Embryology was a point of special pride for Darwin. Why are juveniles of a species often so similar to one another but the adults so different? This was the key fact in von Baer's thinking about development. The answer is simply that natural selection operates on adults but has no need to operate on juveniles, cocooned as they are in their mothers' wombs. Again he drew an analogy between the breeder's world and the natural world. "Fanciers select their horses, dogs, and pigeons, for breeding, when they are nearly grown up: they are indifferent whether the desired qualities and structures have been acquired earlier or later in life, if the full-grown animal possesses them."[5]

As Darwin came to the end of his treatise, he was explicit about the deistic stand he was taking. "Authors of the highest eminence seem to be fully satisfied with the view that each species has been independently created," but this was not Darwin's position. "To my mind it accords better with what we know of the laws impressed on matter by the Creator, that the production and extinction of the past and present inhabitants of the world should have been due to secondary causes, like those determining the birth and death of the individual." And so he concluded:

> From the war of nature, from famine and death, the most exalted object which we are capable of conceiving, namely, the production of the higher animals, directly follows. There is a grandeur in this view of life, with its several powers, having been originally breathed into a few forms or into one; and that, whilst this planet has gone cycling on according to the fixed law of gravity, from so simple a beginning endless forms most beautiful and most wonderful have been, and are being, evolved.[6]

The canny Darwin had run with the empiricist hare and hunted with the rationalist hounds. And he never let anyone forget it. He was always ready to defend the analogy with artificial selection and to suggest that the unifying power of his theory was unassailable.

Darwin and Progress

As the grandson of both Erasmus Darwin and Josiah Wedgwood, Charles Darwin was one of the most ardent social progressionists of the nineteenth century.[7] Why would he not be? The Darwin and Wedgwood families—Charles and Emma particularly—were doing incredibly well during the industrial revolution. Family members on both sides could see the products and benefits of technology, finance, and business. More than this, ideologically they were connected with progress, being deists and Unitarians (Emma was an enthusiastic believer). And the people they mixed with were progressionists. When Charles Darwin came back from the *Beagle* voyage, through his older brother Erasmus he got to know (and admire) people like Thomas Carlyle, not to mention Richard Owen, who was at the time pursuing his own personal path of Germanic idealism.

This said, Darwin knew that progress was a dangerous doctrine to embrace, especially in professional scientific circles. Hints of progressivist thinking put people on guard about evolution. On reading *Vestiges*, he warned himself (in a note scribbled on the flyleaf) to stay away from terms like "higher" and "lower." Yet, like everyone else, Darwin certainly thought that evolution leads to biological progress, and he could not—and truly did not want to—deny this fact.

Following earlier biologists in adapting a Christian metaphor for his own use, Darwin's picture of evolution was of a tree of life, reaching ever upward. "The affinities of all the beings of the same class have sometimes been represented by a great tree. I believe this simile largely speaks the truth." The end twigs represent species, and the lower branches are former twigs that have succeeded and grown thick and given birth to many subsequent descendent species. The divisions of branches represent the branching of life as different species take different directions. "As buds give rise by growth to fresh buds, and these, if vigorous, branch out and overtop on all sides many a feebler branch, so by generation I believe it has been with the great Tree of Life, which fill with its dead and broken branches the crust of the earth, and covers the surface with its ever branching and beautiful ramifications."[8] This metaphor is repeated in the famous final passage, with its reference to the "production of higher animals," directly following. And when he dealt with the fossil record, although he was somewhat cagey on the subject, Darwin made clear that he did see it as moving from simple to complex.

But not in a straight line from lower to higher, as on a ladder or great chain of being. For Darwin, evolution produced branches—many species from one parent species, as had happened with all those Galapagos birds. He was very proud of an idea that he called his "principle of divergence," whereby species form and differentiate by exploiting different ecological niches and avoiding competing with one another. This all tied in with his enthusiasm for von Baer's embryology, which also stressed branching rather than simply a unidirectional change. Also in play were factors that we might regard as more philosophical or theological. Natural selection was there to explain design. If God is the designer, then you expect perfection—or at least the best possible. But if natural selection is the designer, you do not need such perfect adaptation—everything becomes relative. Top and bottom are no longer absolute. As Darwin came to realize, what matters is not being perfect but just being better than the next competitor who comes along.

Recognizing this relativism was critical for a person trying to produce a mature professional science, in an area where hitherto the sole occupants had been varieties of pseudoscience. A big factor differentiating real science from pseudoscience is that in the former (unlike the latter) one tries to say how the world is, rather than how one would like it to be. Professional scientists try to rid their work of cultural values, which can color or distort their understanding.[9] Cultural values might include notions like "the right way to behave in a democracy" or "what would Jesus do?" While trying to ignore these "absolute" values, they embrace "relative" values, as when one thing is compared with another. These occur throughout science: "hotter than," "bigger than," "faster than," or, in Darwin's relativized understanding of natural selection, "better at surviving and reproducing than" (in the language of today's evolutionists, "fitter than").

Adaptation and progress are both end-directed processes—teleological—and they are connected. One has to do with the ends of individual organisms, and the other has to do with the ends of life itself, that is, of all individual organisms considered together. Once you relativize adaptation, seeing it as a matter of being better than competitors rather than better on some absolute scale, you are on the way to relativizing biological progress. Rather than seeing life as a march up a chain of being—as Lamarck saw it—you are seeing it as a matter of scrambling above the next person, or the next species or higher taxonomic group (mammals above reptiles, for instance). We have better adaptations—mammals can function in colder climates better than can reptiles—but that does not mean that on some absolute scale mammals are "better" than reptiles. Indeed, professional scientists train themselves not to see even intelligence as a good thing per se but rather as an adaptation that puts its possessors at an advantage above other creatures—not always, certainly, but on average or in the long run.

By the third edition of the *Origin* (1861), Darwin had spotted this relativism in his own work and was using it to explain a kind of progress. He argued that organisms get involved in what today's evolutionists call arms races, where competing lines of organisms improve their adaptations in competition with one another. The prey gets faster and so the predator gets faster. This includes humanlike adaptations, such as intelligence. One senses that Darwin read more out of his arguments than he put in. "If we take as the standard of high organisation, the amount of differentiation and specialisation of the several organs in each being when adult (and this will include the advancement of the brain for intellectual purposes), natural selection clearly leads to-wards this standard: for all physiologists admit that the specialisation of organs, inasmuch as in this state they perform their functions better, is an advantage to each being; and hence the accumulation of variations tending towards specialisation is within the scope of natu-ral selection."[10]

This all sounds a little absolutist—humans are best because they are the most specialized. For the moment, let us just say that whether these speculations are more wishful thinking than solidly proven sci-ence—whether Darwin was too much of a Victorian to grasp the full implications of his own theory—in truth the happy world of inevita-ble progress up an objective scale seems to be gone. A worldview that accepts the full implications of Darwinian natural selection has no place for absolute values, including absolute progress. One might tack progress on in some way—and indeed the tree of life sets history within a progressivist context—but the causal heart of Darwinian theorizing is against the idea of progress.

After the *Origin*

Anybody's life would tend to be a bit anticlimactic after producing one of the great scientific works of all time, and this certainly holds true of Darwin. Yet, he was to live another twenty-three years, contin-uing in his role as the recluse of Downe, and working hard at the things that interested him right up until his final days. Darwin became particularly intrigued by plants—ideal subjects for one in his posi-tion—beginning with a spritely little book on orchids which pointed out their many and intricate adaptations, especially for fertilization. But Darwin had unfinished business left over from the *Origin*, and he could not let this go.

Homo sapiens, above all, was a topic demanding discussion. In the

Origin, Darwin had been careful not to make humankind the focal point, wanting rather to establish first the general case for evolution. But the status of our species in the evolutionary scheme was not going to go away, and a decade later, in the *Descent of Man,* Darwin turned to a full-scale treatment of the issue. What is particularly significant about this work is not so much that Darwin put us under the umbrella of evolution or invoked natural selection as the chief cause—one would expect him to do this—but rather that he gave a major role to sexual selection in human evolution. This tactic was primarily a response to Wallace, who by then had embraced spiritualism and was arguing that much human evolution—our intelligence, for example—could not be explained by natural causes.[11] Darwin, to the contrary, argued that even if natural selection could not do the job alone, distinctive human features were possibly explicable through the workings of sexual selection. Humans, with their supple brains, chose mates for such things as beauty and intelligence, and thus over the years modern *Homo sapiens* emerged.

By the time that the *Descent* was published, Darwin saw his theory as liberating people from religious superstition. Although he worried about the issue, up through the *Origin* Darwin had not seen natural selection as excluding a designing God. Rather, his theory had made such a God work through the agency of unbroken law—which of course was precisely what the deist God always did. "I see no necessity in the belief that the eye was expressly designed. On the other hand I cannot anyhow be contented to view this wonderful universe & especially the nature of man, & to conclude that everything is the result of brute force. I am inclined to look at everything as resulting from designed laws, with the details, whether good or bad, left to the working out of what we may call chance."[12]

But in the last two decades of his life, Darwin's religious belief faded into agnosticism. He never relinquished his belief in the design-like nature of the organic world, but he saw this as part of the natural order. Religion as such was, for Darwin, mere superstition, but superstition with the virtue of supporting morality. However, morality itself was an outcome of selection for reciprocity—"You scratch my back and I'll scratch yours"—rather than any divinely given command.

Although Darwin could hardly get away from the religion issue, it was not really the main focus of his concerns. He was a fulltime scientist, and what he wanted to do was produce a theory of evolution that would measure up to the exacting standards of the best science in his day, namely, physics and chemistry. These were the standards he had learned in his youth, and these were the standards that ruled throughout his life. No longer was evolution to be a pseudoscience, a marionette pulled by the strings of ideology, but rather a full, working, professional science. As a person, Darwin was attracted to some form of evolutionism, but as a professional scientist, his aim was to produce a scientific theory of evolution. How far Darwin succeeded in his aim is our next question.

5

Failure of a Professional Science

Now it befell that, on the very shore, and over the very rocks, where Tom was sitting with his friend the lobster, there walked one day the little white lady, Ellie herself, and with her a very wise man indeed Professor Ptthmllnsprts . . . He held very strange theories about a good many things. He had even got up once at the British Association, and declared that apes had hippopotamus majors in their brains just as men have. Which was a shocking thing to say; for, if it were so, what would become of the faith, hope, and charity of immortal millions? You may think that there are other more important differences between you and an ape, such as being able to speak, and make machines, and know right from wrong, and say your prayers, and other little matters of that kind; but that is a child's fancy, my dear. Nothing is to be depended on but the great hippopotamus test. If you have a hippopotamus major in your brain, you are no ape, though you had four hands, no feet, and were more apish than the apes of all aperies. But if a hippopotamus major is ever discovered in one single ape's brain, nothing will save your great-great-great-great-great-great-great-great-great-great-greater-greatest-grandmother from having been an ape too.

*C*harles Kingsley was a man of many sides and talents. He was an Anglican clergyman, sufficiently successful to be a chaplain to the Queen. He was for a time professor of modern history at the University of Cambridge. He was a controversialist, having taken on John Henry Newman over the honesty of Roman priests, an attack that spurred Newman to write his *Apologia Pro Vita Sua* in his own defense.

Kingsley was one of a number of religious thinkers, labeled "muscular Christians," who tried to revitalize the Church of England, especially in its approach to the lower classes. He was a gifted amateur artist, whose very rude drawings of a naked, crucified Saint Anne, which he sent to his future wife, Fanny, were as remarkable as was the reaction of his fiancée, who found them to be an acceptable gift. And he was a novelist, especially of stirring historical romances for boys.

His best-known work, *The Water Babies*, a kind of fairy tale, tells of Tom, a filthy and much-abused young chimneysweep who mistakenly comes down into the room of a little girl called Ellie and then runs away and becomes a water baby—a funny little amphibious creature with gills, and very clean. As we follow Tom through his travels and travails, Kingsley, a keen naturalist, took the opportunity to introduce topical jokes and parodies, and among the people at whom he poked fun was Thomas Henry Huxley, otherwise known as Professor Ptthmllnsprts ("Put them all in spirits"). The debate to which Kingsley referred took place at the annual meeting of the British Association for the Advancement of Science in Oxford in 1860, between Huxley and Richard Owen, over the putative status of humankind. Owen had claimed that we humans are different from other apes because we uniquely have in our brains a hippocampus major, and Huxley flatly denied this. In Kingsley's hands, and forever more in people's minds, the difference was translated into one of possession (or otherwise) of large African mammals in our brains. We humans have them. Are the apes similarly privileged? The debate over Darwinism had begun.

No sooner had Darwin spoken than evolution (especially in Britain) started to become fashionable. Although some readers publicly opposed the theory, and the very old guard died off without ever accepting evolution as fact, for many people becoming an evolutionist was an overnight conversion experience.[1] The newspapers and magazines of the day bear this out, as do the examinations students were expected to take for the new science degrees at Oxford and Cambridge. Whereas previously students had been required to show why evolution is false, by 1865 the truth of evolution was assumed and students were expected to be able to discuss its causes.

Evolution had been in the air for a long time, and many people had been looking for a reason to believe it. Darwin's *Origin* gave them such a reason. The work was carefully argued and drew on a massive amount of observational evidence and already-existing theory. Moreover, it came from Charles Darwin, a man known and respected as the author of one of the best travel books of the day, a serious published scientist who battled on despite much sickness, and, above all, a quintessential English gentleman. Father of a large family, devoted husband, rich but not showy, a good master and friend, not given to flashy, foreign displays of intellectual brilliance, just solid and stable, with a few ordinary human failings, including an appalling difficulty with the German language, which he insisted on pronouncing as though it was English—all of these traits endeared Darwin to his fellow countrymen.

But if the fact of evolution was established by the *Origin*, Darwin's mechanism of natural selection fell flat. Virtually no one thought that it could do all that he asked. Almost everyone agreed that selection certainly operated and made for some small-scale, perhaps winnowing, change. But few believed that selection could transform one species into another very different species. People instead opted for a whole range of other putative causes, ranging from Lamarckism (the inheritance of acquired characters), saltationism (evolution by jumps or leaps), guided variation (where the changes may be small but are all in the right direction), orthogenesis (a kind of evolution by momentum, where forces drive whole groups, rather than individuals, along developmental paths), and more. Why was natural selection rejected?

A popular (and sensible) answer is that the scientific evidence failed to convince.

In the first place, natural selection was not critical to acceptance of evolution. A person could become an evolutionist on the basis of homologies alone. In fact, if one were to take a naturalistic approach to the living world, one would have difficulty becoming anything else. But it did not follow that natural selection was the sole (or even major) cause of evolution. Concern with adaptation (that is, design) had been an obsession of naturalists and natural theologians in the 1830s, when Darwin was doing his creative work. But by the 1850s, thanks to men like Owen, Germanic ideas were invading British biology and making it less pressing to explain adaptation above everything else.

More than this, Darwin's theory had its own problems that made selection less than compelling. Two things stand out.[2] The first is the matter of heredity. In the *Origin*, Darwin had no theory about how traits are passed down from one generation to the next, and critics pounced on that deficiency. A Scottish engineer, Fleeming Jenkin, pointed out how difficult it would be for any new variation, however favorable, to become established in a new population. In a very Victorian sort of analogy, he argued that a white man stranded on an island populated by dark savages would hardly turn the island's population white, no matter how successful he was in producing offspring. The same was true in nature. To answer this and similar objections, in a later work Darwin devised a theory he called pangenesis, involving little gemmules being given off all over the body and being collected in the sex cells. But no one was much convinced. Darwin's own cousin, Francis Galton, a respectable scientist himself, dutifully pointed out that there was no evidence of such gemmules or of their being carried by the blood or other bodily substances.

The other big problem facing the theory of evolution through natural selection came from the physicists. William Thomson (later

Lord Kelvin) argued that the earth is much too young to accommo-
date so slow a process as natural selection.[3] Working with evidence
such as the cooling of the earth and the salinity of the seas, the physicists argued that the earth's age falls between 25 and 400 million
years, with about 100 million as the most likely estimate. And since
for much of that time the earth would have been too hot to allow life
to develop, the actual time available for evolutionary change was much
shorter still.

Darwin himself wriggled uncomfortably in the face of this seemingly unanswerable argument. He tried speeding evolution up somewhat, especially by an increased use of Lamarckism (a subsidiary
mechanism he had always accepted), and others chipped in to help.
Wallace thought that frequent ice ages intensified the struggle for existence by increasing selection pressures, and this would have cut
down the time needed for adaptive change. But essentially Darwin just
had to tough it out and hope that the physicists were simply wrong.
Which of course, as we now know, they were. Their calculations were
off in part because they were ignorant of radioactive decay, which
kept the earth warm and slowed the cooling process. Physicists today
estimate that the earth is approximately 4.5 billion years old—quite
enough time for a leisurely evolutionary mechanism such as natural selection to operate. Sadly, Darwin died without ever knowing of this
resolution.

Although these are surely good scientific reasons why natural selection did not become the causal center of a professional school of
evolutionism, they cannot be the whole answer. The fact is that, whatever the scientific problems, natural selection could have been widely
accepted by the public, and it could have provided the basis for a respectable research program. As soon as the *Origin* was published,
Henry Walter Bates (shortly to be followed by Wallace) did some
stunning work which demonstrated that selection could solve a knotty

biological problem, namely, mimicry in moths and butterflies. He showed in great detail that some species of these insects were physically very similar to other quite unrelated species of these insects, and this similarity was a direct function of natural selection. When mimicry occurred, it was always the case that one tasty group of moths or butterflies mimicked another group that tasted nasty. The nasty-tasting insects had built-in protection against predators, especially birds, which avoided them. To the extent that the tasty insects were successful at mimicking the coloration and appearance of the foul-tasting bugs, they escaped predation as well, because the birds mistakenly thought they would also be unpleasant to eat.

Thus, the birds provided the selection pressure that favored mimicry. By eating individuals that happened to look less like the nasty-tasting insects, they increased the chances that the better mimics would survive and produce offspring that would look like themselves. Following through from his observations, Bates ran simple experiments showing unambiguously that the predators had no objection to the taste of the mimics. They were simply fooled into thinking that the mimics were other than they really were.[4]

Darwin himself was highly appreciative of this work, praising it to the skies. Through his publisher, John Murray, he got a book by Bates published and then he found Bates fulltime work as secretary to the Royal Geographical Society. But this very fact reveals both the positive and negative side of the story. The kind of research that Darwin envisioned could be done. A professional science could surely have been started which looked not only at issues in the wild but also studied variation in the laboratory, by, for example, breeding generations of organisms to the point where reproductive barriers between the earlier and later variations arose. Not all of these experiments would have worked, but something would have succeeded and taken the issues a step further.

This was not to be, however, and Bates—the one man who really did do innovative research on natural selection—became a servant of geographers, unable to focus much at all on his own science after the mimicry studies.

Thomas Henry Huxley

So what did happen in the life sciences after the *Origin*? To answer this, we have to look at the aims and achievements of Darwin's supporters, especially Thomas Henry Huxley.[5] Here we have a man whose background was very different from Darwin's. His father, who drifted in and out of madness, was a schoolteacher at the bottom of the middle classes, and young Tom made his way up by hard work and brilliance, graduating as a medical doctor and then, at the end of the 1840s, becoming ship's surgeon on *HMS Rattlesnake* during its voyage to the South Seas. But whereas Darwin had dined with the captain of the *Beagle*, Huxley took his suppers with midshipmen, the boys training to become officers.

On this productive trip Huxley laid the foundation of his career as a morphologist, doing wonderful comparative analyses of jellyfish that he, uniquely, was able to dredge up. He admitted that he never had a feel for naturalism as such—for investigating, for example, the life histories of animals in the wild as they struggled to elude predators, capture prey, and attract mates. He was always interested first, foremost, and finally in comparisons of form and structure from one organism with another. From the beginning, he was much influenced by the German transcendentalists, particularly von Baer. In the 1850s Huxley rose rapidly to the front rank of British morphologists, becoming a professor at the Royal School of Mines in London and braving the enmity of Owen (who could not have enjoyed such a brash young rival).

Huxley intended, from the start, to be a professional biologist. This led him to become an outspoken critic of evolution prior to the publication of the *Origin*. In 1854 he wrote a savage review of Robert Chambers's *Vestiges*. But by the end of the decade he had swung around to being Darwin's most enthusiastic proponent. Like Saint Paul, another notable convert to a great idea, Huxley became something of a fanatic on the subject, and for the rest of his life lectured to audiences of scientists and laypeople alike on evolutionary matters.

The analogy with Saint Paul, which was noted by Huxley's contemporaries, rewards further scrutiny. Paul took on the job of picking up where Jesus left off and making a religion from his master's teachings. Likewise with Huxley and Darwin. However much he would have welcomed it, Darwin himself did not have the ability to found and promote a discipline of evolutionary biology—a network of professional scientists focusing on selection studies. Partly he lacked this ability because he was sick, but there was more to it than this. Darwin had the obsessive, self-centered drive of a truly creative person. His life revolved around his own intellectual needs and interests, and he could never tear himself away from these to go out and do the boring and sometimes dirty work of idea promotion. Perhaps, being a child of privilege, he had been given too much, too early, too easily. Darwin certainly lacked the psychological toughness to do battle in the public arena. He never learned the brutal skills that empire builders like Huxley possessed.

Hence, Darwin left the hard work of discipline-building to his followers, especially Huxley. And here the Saint Paul–Thomas Huxley analogy becomes particularly pertinent and ironic. The religion that Paul founded was successful and viable, but whether it was properly the religion of Jesus is another question entirely. Jesus was a visionary who did not leave a set of prescriptions for everyday life. His teaching was apocalyptic: thinking the end was nigh, he exhorted his followers to give up everything—their money, friends, and family—to follow him. Paul's task was to rein in this millennial enthusiasm, give people rules for everyday living, and generally set up a system of belief and practice that could function and endure here on earth. He had to make sense of Jesus's teaching for folks whose most pressing concerns were not being miraculously lifted up to heaven but rather earning their keep, being spouses and parents, dealing with masters and servants and slaves, and much more. So likewise with Huxley. Darwin may have had the vision, but Huxley had the onerous job of doing something with it, and like Paul he did with it what fit into the world as he saw it, and as he could and would change it.

After Jesus, Paul became the key figure in Christianity. And after Darwin, Huxley became the key figure in evolution. Huxley was not just a man of abilities but a man with tasks and visions of his own. Around the time the *Origin* was published, as Huxley was coming fully into his powers and authority, Britain was in sad and urgent need of reform.[6] It had suffered a huge loss of life during the Irish famines of the 1840s and the disastrous Crimean campaign in the 1850s, followed in short order by the horrific Indian Mutiny, during which women and children were locked up and murdered. It became apparent to many, particularly those who were not part of the old aristocracy of landowners, that Britain was in need of help. No longer a rural-based economy, Britain had become a nation of urban centers, and it needed a professional government that could respond to this dramatic change. Consider the growth of towns, perhaps the most important fact to be presented in this tale. Between 1831 and 1851, London jumped from 1,900,000 to 2,600,000 inhabitants, Manchester from 182,000 to 303,000, Leeds from 123,000 to 172,000, Birmingham from 144,000 to 233,000, and Glasgow from 202,000 to 345,000. In just fifty years, between 1801 and 1851, Bradford grew from 13,000 to 104,000 souls.[7]

At the most basic level, a burgeoning population needs more schools, a larger police force, an extensive postal service, city planning to provide affordable housing, some kind of health-care system, and sanitation engineers. Think of the last of these. In a village, you can open the front door every morning and empty the chamber pot. In a town the size of Bradford, imagine the consequences by mid-century if everyone did that. Tom-All-Alone's, the London slum in *Bleak House*, gives an idea of what such a life would be like. But proper plumbing and sewers come only when governments have the authority to commission planners and engineers, and access to funds to pay for the work. These new urban experts—the planners and engineers—must be trained, which means that schools of higher education must be established and funded. We are not now talking about studying the classics at Oxbridge but studying technology and science at regular-type universities and polytechnic institutes. Here, in education, was the place for Huxley to exercise his talents, and he did. He worked hard to upgrade and professionalize his own field of biology, starting from his base at London University, which was growing at a tremendous rate, and expanding into the science university founded in South Kensington on profits from the Great Exhibition of 1851.

Huxley became an exceptional administrator. He understood full well that in building an academic discipline, the bottom line was who pays? One could not set out to build a business without customers, and in education the customers were students, who in turn provided funds for teachers and researchers. So Huxley's first task was to entice students into biology. One specialty where Huxley and his colleagues were particularly successful was physiology. Medical professionals had begun to realize that the time had come to stop killing patients and start curing them, and they were desperate for new entrants with proper training. Huxley was happy to provide a steady stream of graduate physiologists ready for medical school, and the medical profession was overjoyed to have such a supply.

In reforming education within his own field of morphology, Huxley argued that the hands-on experience of cutting up dogfish and rabbits in university laboratories would be far better training for modern students than reading Plato or Aristotle. His most famous comments on this subject were made in a letter to Kingsley: "Science seems to me to teach in the highest and strongest manner the great truth which is embodied in the Christian conception of entire surrender to the will of God. Sit down before the fact as a little child, be prepared to give up every preconceived notion, follow humbly wherever and to whatever abysses nature leads, or you shall learn nothing."[8] Empirical experience was the key to modern training, and Huxley meant to see that students got it. To this end, he ran for and sat on the new London School Board, which was concerned with primary education in the capital city. He also founded summer training courses for teachers in South Kensington. His best-known biology student was the future novelist H. G. Wells.

Evolution as Popular Science

Now what about evolution in the midst of all this reform? Not only could Huxley not see how to make evolution into a functioning science akin to morphology and physiology, he did not particularly *want* to do so. He could not see a way to find paying customers to support its evolutionary research. It would not cure a pain in the belly, and its taint of unorthodoxy and scandal made it politically unsuitable as instruction for young school children. More than this, evolution was a nuisance to Huxley the professional morphologist. Changes in species concealed relationships and wrecked isomorphisms. Natural selection in particular was far more trouble than it was worth to him. No one could see it in action, changing species, and adaptation had always been the bane of the anatomist, who needed to discern true homologies, not superficial contrivances.

Fascinating (and very revealing) is the fact that Huxley, the archevolutionist, taught virtually no evolution to his own students. He was a brilliant teacher, whose lectures spread over two years and required over a hundred and fifty classes (not to mention practica where one dissected specimens). Yet during this course, which was a marvel of detail and instruction, evolution was lucky to get half a lecture, and natural selection five minutes! This fact amazed not a few students themselves. "One day when I was talking to him, our conversation turned upon evolution. 'There is one thing about you I cannot understand,' I said, 'and I should like a word in explanation. For several months now I have been attending your course, and I have never heard you mention evolution, while in your public lectures everywhere you openly proclaim yourself an evolutionist.'" The answer is that evolution just did not fit into Huxley's vision of science education. In class, he had time "to put the facts fully before a trained audience. In my public lectures I am obliged to pass rapidly over the facts, and I put forward my personal convictions."[9]

But Huxley could see a place for evolution in the larger world of affairs, and here the Saint Paul analogy really comes into play. Huxley and his fellow reformers rightly saw the Church of England as allied with all of the reactionary forces against which he was fighting. He realized that he needed a rival ideology, a kind of secular religion, that he could use to fight traditional Anglican Christianity. Better to have a positive replacement than simply negative arguments. Evolution fit perfectly into this slot. It gave answers to cosmic questions, including where did we come from? It reserved a special place for humankind, especially among those, including Huxley, who interpreted evolution as a progressive process. And it offered much else, all in a materialistic and agnostic (a word invented by Huxley himself) fashion. Natural selection, with its whiff of natural theology, was if anything a detraction, so Huxley minimized or ignored it in his public lectures, but he championed evolution.

Darwin grumbled that Huxley's lectures to working men distorted his theory. But this misunderstood Huxley's intention, which was not so much to push evolution as to push evolution with a meaning. Darwin had alluded just briefly to humans in the *Origin*, but by the early 1860s Huxley had written an entire book on the subject, *Man's Place in Nature*. It started off with a bang in the frontispiece, which showed unambiguously how humans evolved from, and triumphed over, their brutish ancestors. In the *Origin* Darwin was trying to rescue evolution from a Chambers-type evolutionism and to establish a theory of evolution in its own right, separate from an ideology or quasi-religion. Huxley tacked the "ism" right back on again.

The most famous story in the whole Darwinian saga is of Huxley squaring off against the bishop of Oxford, Samuel Wilberforce, at that same 1860 meeting where he had clashed with Owen over the human brain. Supposedly, Wilberforce asked Huxley if he was descended from monkeys on his grandfather's side or his grandmother's side. Supposedly, Huxley replied that he had rather be descended from a monkey than from a bishop of the Church of England. In fact, the encounter was probably very different from that of popular lore. The bishop and his supporters were well satisfied with their showing, and my suspicion is that everyone went off happily together for supper—a good fight and a good meal, with lots to drink.

But rather like Moses leading the Jews to freedom, the precise historical facts were not really the key issue here. What mattered is that the Huxley–Wilberforce debate took on a mythic life of its own, representing the clash between the reactionary forces of Christianity and the reformative forces of Darwinism. One had to choose between looking back or looking forward, and Huxley exhorted the public to look forward. Even the cartoons of the main antagonists in the popular magazine *Vanity Fair* underlined the symbolic issues. The bishop in his garb looked like someone out of the Elizabethan Reformation, while Huxley in his business clothes looked like a Victorian industrialist.

Professional Evolutionary Thinking

Evolution was no longer just a pseudoscience. The *Origin* had seen to that. Although Huxley was no less an evolutionist ideologue than Chambers, evolution itself, as a scientific theory, had certainly moved forward. Huxley later regretted the intemperance of his review of *Vestiges*, but he never regretted his judgment about the book's weaknesses. After Darwin, the status of evolutionary thought (about the fact of evolution, that is) had been jacked up from a pseudoscience to the level of what one might call acceptably popular science, and this put Huxley in a position to formulate a respectable evolutionism—that is, a science-based ideology that could explain the meaning of life. This had not been the case before the *Origin.*

But was there a professional science of evolution after the *Origin?* Certainly not the kind of science that Darwin had hoped for—university based, with natural selection at the center of investigations, as both the tool of inquiry and the explanatory foundation. But it would be wrong to say that no professional evolutionary science existed at all. What existed came from Germany rather than from England, however, and—for all that he labeled it "Darwinismus"—the chief architect was the neo-*Naturphilosoph* Ernst Haeckel.

Brought up on morphology and archetypes, Haeckel seized the idea of evolution with enthusiasm. Downplaying the mechanism of selection, he argued for a thoroughly materialistic interpretation of life's history, where underlying forms are linked to their ancestors and work their way upward until they reach the apotheosis of the tree, humankind. Blended in were the twin convictions that the key to understanding organic relationships lay in the developing organism rather than the adult, and that significant parallels existed between the developing organism and the developing group or race. Haeckel gave all of this an evolutionary interpretation in his so-called biogenetic law: ontogeny (the development of an individual) recapitulated phylogeny (the development of the group).[10]

This kind of thinking resonated in Britain with Huxley and his students, steeped as they were in Germanic transcendentalist thought and keen as they were to build universities on the German model rather than Oxbridge. It found favor in North America also. Darwin's great champion in America was the Harvard botanist Asa Gray, to whom he gave a sneak preview of his manuscript in 1857. Gray immediately converted to evolution.[11] But being a Presbyterian, he could never bring himself to a completely naturalistic position, always believing that the variations on which selection operated had to be guided from above.

Gray's championing of Darwin led to an open fight with the leading American biologist of the day, Louis Agassiz. A Swiss-transplanted ichthyologist, discoverer of ice ages, Harvard professor, and now founder of a massive museum of comparative zoology, Agassiz was a very public figure with the enviable ability to raise huge sums of money. He moved in Boston intellectual circles that included Ralph Waldo Emerson and Henry Wadsworth Longfellow, and he taught many successful students.[12] To the day of his death in 1873, Agassiz was also an arch opponent of evolution, and he debated the issue publicly with Gray in the years after the *Origin*'s publication.

History has judged the evolutionist Gray as the winner of those debates. But Gray seems to have won a battle rather than the war. Agassiz, a dyed-in-the-wool *Naturphilosoph,* had the real and lasting influence on evolutionism in America. He had studied with two of *Naturphilosophie's* leaders, the philosopher Friedrich Schelling and the biologist Lorenz Oken, and he saw a three-fold parallelism between the history of life, the history of the individual, and the systematic array of living organisms. "One may consider it as henceforth proved that the embryo of the fish during its development, the class of fishes as it at present exists in its numerous families, and the type of fish in its planetary history, exhibit analogous phases through which one may follow the same creative thought like a guiding thread in the study of the connection between organized beings."[13] Although Agassiz himself interpreted these in a nonevolutionary fashion, his vision of life's history laid itself open to an evolutionary interpretation.

The next (post-*Origin*) American generation had no great qualms about breaking entirely with miraculous creations and happily turning evolutionist. If it happened more slowly in America than in Britain, that was because the U.S. Civil War was being fought in the years right after the *Origin's* appearance, but gradually and definitely evolution came to America. Like Huxley's in England and Haeckel's in Germany, America's brand of evolution had a transcendentalist flavor and owed little to the writings of Charles Darwin himself. Rather than being a working hypothesis that could be used to investigate the questions of the life scientist, evolution in Cambridge, Massachusetts, was, for the most part, a metaphysical belief that had more in common with Munich, where Agassiz was trained, than with Cambridge, England, where Darwin was trained.

Often it was difficult to tell from one publication to the next when an American biologist had changed from a creation-based view of organisms to an evolution-based view. To see how strange this is, imagine not being able to tell whether a molecular biologist had read Watson and Crick's 1953 paper explaining the structure of the DNA molecule as a double helix. While the Americans should not be denied the status of professional scientists, they hardly deserved the highest marks for the work they produced. It was not just that American evolutionists tended toward archetypes and homologies—form rather than function—but also, as with Huxley and the English morphologists, they had little genuine or fruitful interest in causal issues.

Compounding and reinforcing this lack of interest was a situation peculiar to the New World—its huge fossil deposits. As the American West was opened by the railroads in the years after the Civil War, discoveries of prehistoric fossils were made almost on a daily basis. From Colorado, Montana, Wyoming, and other states came reports of fantastical monsters, reptiles of truly gigantic size and shape, buried in the rocks but already poking out into the air. Rich men sent out teams to excavate, and specimens began arriving back East in short order. Digging was frenetic, and although many specimens were lost through crude and amateur methods of recovery and transportation, the overall results were magnificent: allosaurus, ceratosaurus, brontosaurus, camarasaurus, amphicoelus, diplodocus, camptosaurus, and many, many others. Everyone learned about the stegosaurus, that peculiar creature looking like a python on legs that has swallowed a goat and has ridiculous plates sticking out all along its back. It gave the cartoonist Gary Larson nonstop material, and led to much speculation about the animal's way of thinking, for its brain was incredibly small and there was a kind of subsidiary brain or nerve center down at the other end.

The creature had two sets of brains—
One in his head (the usual place),

The other at his spinal base.
Thus he could reason *a priori*
As well as *a posteriori.*
No problem bothered him a bit
He made both head and tail of it.

As a result of these rich fossil finds, paleontology became the big biological science in America, and paleontology was simply not a science where experimentation was available or where functional issues predominated.

In 1876 Huxley visited the New World to give public lectures in New York. In search of dramatic examples to illustrate evolution, he settled on the horse. On a visit to Yale University, where the collector Othniel Charles Marsh had a major collection, Huxley was quite staggered by the equine materials. At once, he revised his lecture to state unequivocally that the horse, although extinct in North America by the time humans arrived there, had evolved in America and not in Europe. The result was a famous and widely reproduced picture of the lineage of the horse, from the four-toed ancestor to the single-toed living representative.

Moreover, the picture came with a prediction: "In still older forms, the series of the digits will be more and more complete, until we come to the five-toed animals, in which, if the doctrine of evolution is well founded, the whole series must have taken its origin."[14] Within two months, Marsh discovered just such an animal, the famous five-toed eohippus. Today, we know that the horse record is far, far more complicated than Huxley implied, with masses of branches and extinctions. But the forceful simplicity of Huxley's demonstration was precisely the thing that impressed his audience. This was evolution that people understood—evolution that you could really see and grasp.

Museums

Huxley's brand of showmanship was not what Darwin had envisioned as the basis for a professional, selection-based, evolutionary science. Causes really did not matter that much to Huxley, Marsh too was indifferent to them, and the other Americans went off into fantasies of Lamarckism and spirit-type forces. This is not to say that the post-Darwinian evolutionists exhibited no talent or ability. Some triumphs requiring subtle uses of morphology and embryology were registered. In England, for example, Huxley's protégé E. Ray Lankester showed that the horseshoe crab was not really a crab but an arachnid—that it belongs to the family that includes the spiders![15] Hardly evident on the surface of things. Likewise in America, Marsh's great rival, Edward Drinker Cope, was a genius at making inferences about whole organisms from just a few scattered, broken bones. But overall, the stuff of great professional science was lacking. At a time when physics was about to undergo a great paradigm shift, with relativity theory and quantum mechanics, evolutionary biologists pulled back from serious discussion of causes and spent their time spinning fabulous histories of the past.

Evolutionary biology as a professional science was distinctly second-rate. It failed to be properly causal; its "laws" often failed to predict; and worst of all it was riddled with cultural values, especially related to notions of progress.[16] Deservedly, evolution was pushed out of the universities. Thanks to the *Origin*, evolution was more than just a pseudoscience, but as a popular science it was still hitched strongly to extrascientific values. Unlike the popular science of Laplace, who started with professional science and then simplified it, adding extrascientific material as necessary to make the point, evolution was a popular science that had pushed its way up from a pseudoscience, but with many extrascientific values still attached. The popular science of evolution had not yet broken free of the ideology of evolutionism.

Having earned no secure and widely accepted place within the universities, evolution became a creature of the public domain, and it was there that new homes for this great enterprise were found. Natural history museums built in London and New York City became cathedrals or temples for instructing the masses in the tenets of evolutionism. Lankester was appointed director of the British Museum (Natural History) in South Kensington, while Henry Fairfield Osborn, a wealthy American and another student of Huxley, became director of the American Museum of Natural History in New York. In these institutions, an evolutionist could do serious classification studies during the morning hours and in the afternoon supervise the fabulous dinosaur reconstructions that still delight and amaze the public to this day. Generations of schoolchildren were shipped in to learn about nutrition, cleanliness—and about evolution. In the process they got a large dose of instruction in the ideology of progress. The wealthy men who put up cash for the museums had their own ends in view, and funding a value-free enterprise was not one of them.[17] The ideology of progress was running as rampant as it ever had.

6

Social Darwinism

Kim smoked slowly, revolving the business, so far as he understood it, in his quick mind.

"Then thou goest forth to follow the strangers?"

"No. To meet them. They are coming in to Simla to send down their horns and heads to be dressed at Calcutta. They are exclusively sporting gentlemen, and they are allowed special faceelities by the Government. Of course, we always do that. It is our British pride."

"Then what is to fear from them?"

"By Jove, they are not black people. I can do all sorts of things with black people, of course. They are Russians, and highly unscrupulous people. I—I do not want to consort with them without a witness."

"Will they kill thee?"

"Oah, thatt is nothing. I am good enough Herbert Spencerian, I trust, to meet little thing like death, which is all in my fate, you know. But—but they may beat me."

*K*imball O'Hara, the hero of Rudyard Kipling's *Kim* (the greatest novel written by a Westerner about India), is a poor white who is enrolled in the British Secret Service, playing the "Great Game," the battle of wits between the English and the Russians up on the North West Frontier. Assisting him is a greasy fat Hindu who, despite his fears and his appalling half-education, turns out to be the bravest and brightest agent of them all. Desperate to be regarded as a man of intellect and learning, toward the desired end of being elected a fellow of the Royal Society of London, the babu (an Indian clerk) Hurree Chander Mookerjee has contributed countless "rejected notes" on ethology to the *Asiatic Quarterly Review*. And he is up on the latest and best philosophy also, being an enthusiastic follower of Herbert Spencer, as were millions of others around the globe in the late nineteenth century.

Who was this guru, Herbert Spencer, and why did he cast such a spell? Like Charles Darwin, Spencer was born into the middle classes in the British Midlands, although lower on the scale. His family came from the ranks of small shopkeepers, teachers, and the like, who were given to dissent and strong distrust both of the aristocratic elites above and the working classes amassed below. Spencer was wary of the state and all that it represented, and hardly more enthusiastic about unions and other organizations. In other words, he came from a libertarian, laissez-faire, free trade background and regarded the state and its authority—including the established church—with suspicion.[1] He breathed the same atmosphere in which, a hundred years later, the future prime minister of Britain, Margaret Thatcher, was raised. Her father, too, was a Methodist shopkeeper who despised the power of those above almost as much as he feared the threat of those below.

Always independent and eccentric—he never married and lived in boarding houses chosen deliberately for their dullness, so that he would not be distracted from his great projects—Herbert Spencer had overwhelming confidence in his abilities and insights, reinforced by his inability to read books with which he disagreed. (They gave him headaches.) Some training in mathematics fitted him to be a surveyor, a popular job in the 1840s as the railways expanded, and through this occupation (which involved significant earthworks) he encountered Lyell's *Principles*, the reading of which turned him into an evolutionist. He switched to journalism and from then on lived by his pen, bolstered by countless admirers ever ready to cater to his needs and demands—which were many, in food, entertainment, and above all adoration. Psychology, biology, sociology, philosophy—nothing escaped Spencer's gaze or eluded his world picture, which he presented in many, many volumes through the second half of the nineteenth century, describing, explaining, and fleshing out his so-called synthetic philosophy.

In a series of articles through the 1850s, Spencer established his credentials as an evolutionist, and just after the *Origin* appeared he summed up his views in a work modestly described as *First Principles*. Drawing on Kant's notion of the thing-in-itself, the noumenal world that supposedly lies behind all reality, Spencer (with a Germanic flourish of capital letters) spoke of the Absolute or the Unknowable. Both science and religion, properly understood, point to this, and as we appreciate it in our lives and understanding, it reveals itself as being in constant motion—not just motion but evolution (Spencer popularized this term), and evolution ever upward in a progressive mode.

Spencer had read von Baer on embryology, as well as Adam Smith and others on division of labor, and, combining insights from all, he saw progress as being a move from the undifferentiated to the differentiated, or (in his words) from the homogeneous to the heterogeneous:

Now we propose in the first place to show, that this law of organic progress is the law of all progress. Whether it be in the development of the Earth, in the development of Life upon its surface, in the development of Society, of Government, of Manufactures, of Commerce, of Language, Literature, Science, Art, this same evolution of the simple into the complex, through successive differentiations, holds throughout. From the earliest traceable cosmical changes down to the latest results of civilization, we shall find that the transformation of the homogeneous into the heterogeneous, is that in which Progress essentially consists.[2]

Everything obeys this law. Humans are more heterogeneous (complex) than other animals, Europeans more heterogeneous than savages, and the English language is more heterogeneous than the tongues of other peoples. Spencer's law had empirical support, but it was more an *a priori* truth about the world, for it follows directly from the nature of causation. Causes are always exploding outward in the effects that they produce—one cause has multiple effects, and so homogeneity produces heterogeneity. This truth is also mixed up with laws of physics, for the second law of thermodynamics shows that everything tends toward equilibrium, but a moving (or, as it came to be known, a dynamic) equilibrium. A system such as a species is a kind of unstable equilibrium; something disturbs it (any outside cause is going to have uneven effects on a system), and then it moves upward as it strives once again to achieve equilibrium—which it might, only to set off yet again.

Darwin and Spencer were worlds apart. Spencer's picture was inherently progressive in a way that Darwin's was not. This comes through most vividly in their respective attitudes toward natural selection—the mechanism that was so corrosive to notions of absolute progress. For Darwin, selection was always the primary mechanism. Although he admitted other factors, including Lamarckism, they were secondary. Spencer discovered natural selection independently, as it happened, but he always thought of it as unimportant—it got only perfunctory mention in *First Principles*. He put far more emphasis on Lamarck's inheritance of acquired characteristics. Progress counted, not selection. The chief function of Malthusian pressure was to put a premium on effort. Some will strive more effectively than others, and they will therefore develop adaptive characteristics that will then be passed directly to their offspring. Spencer argued that the higher up the evolutionary scale one goes, the greater the brains required for survival, and therefore the less "vital bodily fluid" left over for producing offspring. Eventually, a point is reached where the numbers balance out, and the Malthusian pressures and consequent struggles are over. The greasy pole has been climbed.

Laissez Faire?

Although his philosophy came to be known as social Darwinism, not Spencerism, Herbert Spencer more than anyone else came to epitomize the attempt to draw a moral code for proper living from his beliefs about evolution. The basic pattern of Spencer and all who followed or imitated or worked in parallel with him was simple. The key was the supposed progressiveness of evolution—simple to complex, homogeneous to heterogeneous, blob to Briton. In his view, humans had a good thing going with evolution, and their moral obligation was to help the process along, not to stand in its way. We humans were living proof that when evolution was allowed to do its work, the outcome was positive. So far, so good.

But then the question arose, how does evolution do this work, and for what purpose? Almost all agreed that humans were unambiguously the top species; and for the Victorians, English-speaking Anglo-Saxons were unambiguously the top humans (never mind that sometimes other people compared the Victorians with the loyalty of dogs or the industriousness of insects). Most obviously, the mechanism that put them on top was natural selection, and it seemed an easy transition from the biological mechanism to a socioeconomic policy of laissez faire, where unending struggle and competition was accepted as the natural order because it improves the lives of survivors, while the weakest go to the wall.

Spencer's thinking seems the perfect exemplification of social Darwinism. All those who advocated state-supported amelioration of poverty were compounding the nation's problem, he believed. "Blind to the fact that under the natural order of things, society is constantly excreting its unhealthy, imbecile, slow, vacillating, faithless members, these unthinking, though well-meaning, men advocate an interference which not only stops the purifying process but even increases the vitiation—absolutely encourages the multiplication of the reckless and incompetent by offering them an unfailing provision, and *discourages* the multiplication of the competent and provident by heightening the prospective difficulty of maintaining a family."[3]

But the story is more complex than this passage would suggest. Not only was Spencer no selection enthusiast, but he wrote these sentences before he had written anything about evolution. In other words, he did not make a simple deduction from evolution to morality. With Spencer as with everyone else, science and ethics were a package deal, with the one reinforcing the other and conversely. Rather than a case of cause and effect, Spencer's science and ethics had a common cause, stemming from the same set of convictions. And progress was the backbone of it all. Evolution was derivative; evolution*ism* was basic.

Although struggle was all-important in the process of upward development, in the end it drops away, according to Spencer. Thus, a major component of his world vision was adamant opposition to militarism and especially the arms race between Britain and Germany. He also resisted all kinds of tariffs and other impediments to the free exchange of goods. To quote Spencer's most admiring American supporter, Edward L. Youmans: "The essence of evolution is transformation—the substitution of higher agencies for lower in the unfolding economy of the world. War is certainly one of the things that must certainly be left behind."[4] Again, more than just evolution was at work here, for the Spencer family had Quaker connections and sympathies. Ethics and evolution for Spencer and his acolytes were like hand and glove.

Herbert Spencer was incredibly popular, far more so in his lifetime than Charles Darwin, and his influence was felt around the world. In China, to take but one example, end-of-the-century radical young thinkers were totally absorbed and enthused by evolutionary ideas, especially progress. They loved Darwin, despite his suggestion that our ancestors were apelike. After all, was not he himself showing respect for ancestors by following in the footsteps of his own grandfather, Erasmus Darwin? Yet the true authority for these young Chinese intellectuals was not Darwin but Herbert Spencer, with his philosophy that all is necessarily moving upward and we humans are both cause and effect of this process. "What we can know is that the way of the world must progress, and that what comes will be better than what is." The Spencerian notion of the Unknowable was taken up in the East to support a version of design without a designer—life works its way out and forward, mysteriously. "He who goes with Heaven shall survive; he who goes against it shall perish."[5]

Social Philosophy

Not everyone who used evolution for social or moral purposes was an exclusive or ardent Spencerian. Ernst Haeckel had his own evolutionarily inspired moral system, and Thomas Henry Huxley opposed Spencer toward the end. The important point is that after the *Origin*, evolution provided a foundation or support or cover for many different people who were inclined to think about social and moral issues. The result was evolutionism, rather than just the fact or theory of evolution.[6]

In America, among those who fell in with Spencer's laissez-faire attitudes was the notorious American sociologist William Graham Sumner. "The facts of human life . . . are in many respects hard and stern. It is by strenuous exertion only that each one of us can sustain himself against the destructive forces and the ever recurring needs of life; and the higher the degree to which we seek to carry our development the greater is the proportionate cost of every step."[7] Many post– Civil War businessmen—the so-called robber barons—were also ardent Spencerians, most famously Andrew Carnegie, founder of U.S. Steel. But their relationship to Spencer was often more complex than the simple use of the synthetic philosophy for personal ends.

Apart from anything else, such men generally preferred to stress the positive side of laissez faire—the success of the successful—rather than its downside. In Carnegie's words: "Man was not created with an instinct for his own degradation. From the lower he has risen to the higher forms, and there is no conceivable end to his march to perfection. His face is turned to the light; he stands in the sun and looks upward."[8] Moreover, these men often used their vast riches to promote institutions that would further the upward progress of society, becoming great philanthropists in the process. In Carnegie's case, he used steel money to found, among other things, public libraries, an institute of technology, a teachers' pension fund, an endowment for peace, and other institutes and corporations to aid colleges and universities and promote scientific research. In all, Andrew Carnegie gave away 90 percent of his fortune before he died at the age of 84. This beneficence was very much in the Spencerian mode, for such institutions became places where the poor but bright child—being naturally fitter—could, through self-improvement, rise up in social status.

Christians of Spencer's time were well-practiced at drawing entirely opposite conclusions from one another in answer to the question "What would Jesus have us do?" While some sects championed individualism and self-reliance, others argued that Jesus would have been a socialist, promoting state-sponsored social services and care of the less fortunate. Similarly, evolutionists in the nineteenth century were no less adept at using ideology to promote their own favorite moral and social system, citing either Spencer or Darwin (whose name had more authority, whatever his personal views) to bolster their position. America's Marxists were notorious for using evolution for political ends.[9] In this, they merely followed a tradition started by the founders themselves: "What a sequence of crippled moles, who have chosen shit as their element of life, must be brought to the highest degree of garbage existence by Darwin's *natural selection* before a single *Bettziech* can be produced." Coming from Engels, this, apparently, was an endorsement: "I am just reading Darwin and find him excellent."[10]

Somewhat less scatologically, one of the founding heroes of the British Labour Party, Keir Hardy, likewise used the authority of evolutionists to support his position. "Darwin stated emphatically that 'those communities which included the greatest number of the most *sympathetic* members would flourish best', and in so stating he conceded the whole case for which the Socialist is contending. It is sympathetic association and not individualistic competition which makes for progress and the improvement of the race." Hardy was not out of line: Alfred Russel Wallace, the co-discoverer of natural selection, who named one of his children Herbert Spencer Wallace, was nevertheless a life-long socialist.[11] As a teenager, he had been to hear a speech by the Scottish factory owner and early socialist Robert Owen (no relation to the morphologist Richard Owen), and the experience stayed with him all of his life. A major plank of Wallace's thought was that selection can favor groups, not just individuals, and that these kinds of selection pressure would lead to adaptations that promote group harmony.

This view was shared by others, notably the Russian anarchist Prince Petr Kropotkin. Kropotkin argued (very much in tune with his own countrymen) that the real struggle for existence comes as organisms face the harsh demands of nature, rather than as they compete with their fellow organisms for food or mates. Drawing attention to the appalling weather conditions of his native country—snow, frost, rain, all liable to come at the most inconvenient times—he concluded: "They made me realize at an early date the overwhelming importance in Nature of what Darwin described as 'the natural checks to over-multiplication,' in comparison to the struggle between individuals of the same species for the means of subsistence." The only way that any animal species could survive was by banding together against the elements. "The animal species, in which individual struggle has been reduced to its narrowest limits, and the practice of mutual aid has attained the greatest development, are invariably the most numerous, the most prosperous, and the most open to further progress . . . The unsociable species, on the contrary, are doomed to decay."[12] In Kropotkin's view, animals survive through mutual aid, brought on by a sympathy that exists between organisms of the same species—including human organisms in the human species.

War and Peace

The problem of warfare was of growing interest in the last part of the nineteenth century, and the outbreak of the Great War dominated the beginning of the twentieth century. Many promoters of the naturalness, inevitability, and desirability of conflict based their views on evolution. General Friedrich von Bernhardi, pushed out of the German army because he signaled a little too bluntly the General Staff's intentions, left no place for the imagination in his best-selling *Germany and the Next War* (1912). "War is a biological necessity," and hence "those forms survive which are able to procure themselves the most favourable conditions of life, and to assert themselves in the universal economy of nature. The weaker succumb." In his view, progress depended on war: "Without war, inferior or decaying races would easily choke the growth of healthy budding elements, and a universal decadence would follow." And, anticipating the horrible philosophies of the twentieth century: "Might gives the right to occupy or to conquer. Might is at once the supreme right, and the dispute as to what is right is decided by the arbitrament of war. War gives a biologically just decision, since its decision rests on the very nature of things."[13] As with the socialists, those promoting the necessity of strife in the name of evolution often postulated group selection.

How much the totalitarian philosophies of the twentieth century—nazism, fascism, communism—owed to social Darwinism has been much debated. In the case of Hitler and his gang, historians today dilute any significant role for evolution (Darwinism in particular) and instead put much weight on the influence of cultural factors, such as the apocalyptic anti-Semitism of the Volkish movement which centered around the Wagnerians at Bayreuth.[14] Although one can tie aspects of this movement into the Romantic thinking at the beginning of the nineteenth century, biological connections are remote or nonexistent.

More controversial are those who argue that Christianity itself was the greatest ideological force for ill in Nazi Germany—specifically the anti-Semitic legacy of Martin Luther. As a general thesis, this is overstated, although recent scholarship has shown that connections between Hitler's movement and Protestant Christianity were a lot closer than many had admitted. The fact is, however, that occasionally Hitler blamed Christianity for opposing evolution! Sounding like a paradigmatic social Darwinian, he wrote, "He who wants to live must fight, and he who does not want to fight in this world where eternal struggle is the law of life has no right to exist."[15] But do not read too much into any of this. The theoreticians of National Socialism saw that evolution posed grave threats to the very core of their ideology. Even a progressivist view of evolution that put Germans at the top of the tree would have had to recognize that Jews were on one of the near-by branches. And the monkey origins of Germans as well as everybody else could hardly be concealed. Evolution—as most of the Nazis saw quite clearly—was fundamentally opposed to National Socialist thinking.

On the other side, among the antiwar evolutionists, one had not only Spencer, Wallace, and Kropotkin but also, until it was swallowed up by the First World War, an ardent peace party in America that made much use of evolutionary principles. Fire was fought with fire. Vernon Kellogg, a zoologist at Stanford and popular writer on evolution, pressed his claims that the way of the biological future—the way of progress—led away from conflict and fighting.

> Man is an incident of organic Evolution, and at bottom and in all things his body and his nature are the product of this Genius of Life. And just as Evolution made him, with his need, a Fighter, and taught him War, so now, with the passing of this need, with the substitution of reason and altruism for instinct and egoism, Evolution will make him a Man of peace and goodwill, and will take War from him. And any man will find his greatest advantage and merit in aiding, rather than delaying this beneficence.[16]

Feminism

Gender questions were as contentious for Victorians as they have been for Americans in recent years.[17] Many feminists think that social Darwinism was grotesquely sexist, not just positing differences between men and women but approving them, to the detriment of women.

Darwin's *Descent of Man* in particular contained some conventional Victorian sentiments about the innate qualities and proper roles of the sexes. "Man is more courageous, pugnacious, and energetic than woman, and has a more inventive genius." In compensation, woman has "greater tenderness and less selfishness."[18] Darwin could be describing the relationship between David Copperfield and Agnes Wickfield, and he was most certainly describing his own home life and that of his supporters.

Huxley claimed to be in favor of women's rights, but he still argued that (although much is due to repressive upbringing) women naturally fall behind men when it comes to such things as intellect and drive. And whatever Huxley's theory, his practice always seemed to incline in chauvinist directions. For all that he wrote passionate letters to his fiancée when they were poor and unable to marry, after the event Huxley settled into a very conventional relationship, where he went out and about while his wife and children stayed home—so much so that even friends remarked upon the fact. When Huxley's daughter committed the ultimate liberating move of going abroad and marrying the husband of her dead sister—a union that was not allowed in England, because even entertaining the thought of marrying one's brother-in-law at some future time, after one's sister was dead, was taken as a threat to the sister's marriage—he became very upset.

Victorian delicacy did not preclude mention of a major reason for women's biological inferiority: menstruation. "Even if woman possessed a brain equal to man's—if her intellectual powers were equal to his—the eternal distinction in the physical organisation of the sexes would make the average man in the long run, the mental superior of the average woman. In intellectual labour, man has surpassed, does now, and always will surpass woman, for the obvious reason that nature does not periodically interrupt his thought and application." Or, as Havelock Ellis was to put it, so delicately: "Women are thus, as it

were, periodically wounded in the most sensitive spot in their organism."[19]

Of course, recognition of this basic fact of biology does not in itself demand an evolutionary connection. The common nickname of
the "curse" echoes Genesis more than the *Origin*. But people were
ready to offer causal, evolutionary explanations of female inferiority.
Predictably, although Darwin based his attitude toward women on his
theory of sexual selection, others were happy to turn to alternative,
then-popular ideas, including the biogenetic law. "The higher the animal or plant in the scale of being, the more slowly does it reach its ut-
most capacity of development. Girls are physically and mentally more
precocious than boys. The human female arrives sooner than the male
at maturity, and furnishes one of the strongest arguments against the
alleged equality of the sexes." Little wonder that the boy develops into
a man of intelligence, whereas the girl "has been educated into a
young lady reading novels, working crochet, and going into hysterics
at the sight of a mouse or a spider."[20]

One starts to see, incidentally, where Freud got his ideas about bi-
sexuality and about the tendency of humans to slide to one end of the
scale or the other as they develop—except for those unfortunates who
got stuck in the middle. Boys recapitulate the past, going through a
feminine phase as they mature. Taking time off from his study of fossils, Cope spelled things out clearly. "Perhaps all men can recall a period of youth when they were hero-worshippers—when they felt the
need of a stronger arm, and loved to look up to the powerful friend
who could sympathize with and aid them. This is the 'woman stage'
of character."[21]

Not everyone thought this way, of course. In parallel with discussions of other topics such as war and peace, many ardent evolutionists
drew conclusions directly opposite to those of people (like Darwin
and Huxley) who believed in the innate inequality of the sexes. Given

that Wallace was always one to opt for the uncomfortable and unpop-ular position, it comes as no great surprise to find that he considered feminism and evolution to be soul mates. Apparently, there could be no hope of either biological progress or cultural progress if things were left in the hands of unthinking men. Women must, could, and would come to the rescue.

Through a kind of high-powered sexual selection, the future would see young women making the amorous advances and taking as mates only those young men with the very highest moral and intellectual powers and intentions. Hence, the human race would be able to scale new heights. "In such a reformed society," Wallace wrote, "the vicious man, the man of degraded taste or of feeble intellect, will have little chance of finding a wife, and his bad qualities will die out with him-self. The most perfect and beautiful in body and mind will, on the other hand, be most sought and therefore be most likely to marry early, the less highly endowed later, and the least gifted in any way the latest of all, and this will be the case with both sexes."[22]

Given views like these, one starts to sense why people loved Wallace and yet refused to allow him any position of authority in Victorian Britain. Another century would go by before people started thinking that such ideas might truly reflect human nature, and that perhaps young men should sit back and let young women become the driving force in romantic relationships. Whether that behavioral code was realistic or not, the fact remains that the co-discoverer of natural selection did really think that evolution empowers women and points to their natural superiority.

Decline

As the nineteenth century drew to an end, people began to worry about whether progress was indeed inevitable. Perhaps it had already ended, and now the world was in decline. The biologist E. Ray Lankester was a little obsessive on the subject, forever warning in sepulchral terms about the decay of Britain—in science, in education, in defense, and more, especially compared with Germany. "The traditional history of mankind furnishes us with notable examples of degeneration. High states of civilisation have decayed and given place to low and degenerate states." We should not get complacent. "Possibly we are all drifting, tending to the condition of intellectual Barnacles or Ascidians."[23]

Another Huxley student, H. G. Wells, was also prone to painting scenarios of decay and decline. In his *Time Machine*, published in 1895, Wells gave a frightening foretaste of the possibilities of degeneration. An inventor has created a machine that carries him forward to the year Eight Hundred and Two Thousand, Seven Hundred and One. Here he finds what one of the chapter titles rightly describes as "The Sunset of Mankind." *Homo sapiens* has evolved from our present state into two separate species. There are the Eloi who live above ground and the Morlocks who live beneath the ground. The Eloi are friendly, childlike creatures who seem to spend their whole days playing, without any thought of the future or of the need to provide for themselves or to indulge in any intellectual work of the kind that a Victorian gentleman would find exciting or worthwhile. "A queer thing I soon discovered about my little hosts, and that was their lack of interest. They would come to me with eager cries of astonishment, like children, but like children, they would soon stop examining me, and wander away after some other toy. The dinner and my conversational beginnings ended, I noted for the first time that almost all those who had surrounded me at first were gone."[24]

Then the traveler discovers the terrible truth about the Eloi. They are kept as cattle by their subterranean neighbors. The Morlocks do all the work and provide all the food for those above ground, but eventually they capture them and take them below to be eaten. Just as the Eloi are in a very fundamental sense "degenerates" compared with Victorian humans, so also are the Morlocks, for all their hard-working virtues. "I lit a match, and, looking down, I saw a small, white moving creature, with large bright eyes which regarded me steadfastly as it retreated. It made me shudder." Continuing: "I felt a peculiar shrinking from those pallid bodies. They were just the half-bleached colour of the worms and things one sees preserved in spirit in a zoological museum. And they were filthy cold to the touch." And yet, we have to accept that the Morlocks, no less than Eloi, were descended from us.

> I do not know how long I sat peering down the well. It was not for some time that I could succeed in persuading myself that the thing I had seen was human. But, gradually, the truth dawned on me: that Man had not remained one species, but had differentiated into two distinct animals: that my graceful children of the Upper World were not the sole descendants of our generation, but that this bleached, obscene, nocturnal Thing, which had flashed before me, was also heir to all the ages.[25]

The Time Machine was clearly a clarion call to action: if we want to prevent the decline of the human race, we must do something now. Others were also making and heeding the call. The eugenics movement grew out of this concern. It was an attempt, through persuasion or coercion, to stop some from breeding and encourage others to have large families. So did a peace movement. War leads to the destruction of the finest and fittest, the pacifists argued; the brave men go off to fight and get killed, while the degenerates remain behind and breed. Kellogg was quite desperate on the subject. "War to the biologist seems, above all else, stupid. It is so racially dangerous. It so flies in the face of all that makes for human evolutionary advance, and is so utterly without shadow of serious scientific reason for its maintenance.

It is not natural selection in Man, nor in any way the counterpart of it." He continued. "It does not encourage bravery, but directly and positively robs the race of it. For it kills the brave, and preserves the coward to breed his kind. Hiring willing men to fight, victualling them, transporting them, burying them, is not a stimulus or an exercise of personal hardihood or bravery or human virility."[26]

Those worrying about degeneration were the counterpoint to those cheering on progress. They too thought progress was an evolutionary good, yet they feared that it was not guaranteed and that efforts must be made to sustain or regain it. But was there no one who wanted to get off the merry-go-round entirely, by refusing to see evolution as providing a moral, value-impregnated direction? This seems to have been the path Thomas Henry Huxley (1893) was moving down toward the end of his life. Haunted by the early deaths of his children, wracked by depression and a sense of personal sin, Huxley grew skeptical about the virtues of evolution. The habits of the lion and the tiger did not seem to him to be the epitome of moral character, and by the end he came to think that true morality might lie in combating the force of evolution rather than going with the flow.

But Huxley's was very much a lonely voice in the wilderness. General opinion was that he had lost it. When he died, his gloomy warning all but went with him. The Cambridge philosopher G. E. Moore took up the argument against the evolutionists' moral vision, stressing that the premises did not support the conclusions. But essentially he and like thinkers were ineffective at stemming the tide of progressivist ethicizing.

Secular Religion?

After the *Origin*, a functioning science of evolutionary morphology was in place. It was not very Darwinian, not very high-class, and increasingly unattractive to the best-quality student, but it was professional nevertheless. What about evolutionism, with its progress, moral exhortations, world pictures, and so forth? Is it proper to speak of this as a religion, joining with Hurree Babu in seeing Herbert Spencer as its chief god? "Under the striped umbrella Hurree Babu was straining ear and brain to follow the quick-poured French, and keeping both eyes on a kilta full of maps and documents—an extra-large one with a double red oil-skin cover. He did not wish to steal anything. He only desired to know what to steal, and, incidentally, how to get away when he had stolen it. He thanked all the Gods of Hindustan, and Herbert Spencer, that there remained some valuables to steal."

Of one thing we can be sure. No one now and (more importantly) few back then thought of this kind of evolutionism as a forward-looking, mature science, or of ever having any chance of becoming one. John Tyndall—superintendent of the Royal Institution in London, a friend of Huxley, and a member, with Spencer and others, of the X-Club, a group that met for dinner in the 1860s and 1870s and organized British science while eating their roast beef—was busy articulating the norms and practices of good science. Like Herschel and Whewell a generation earlier, he worked to block out a space in Victorian society for the professional scientist, trained and worthy of employ.

Tyndall stressed above all the need to be objective and to keep personal opinions and motives and desires out of science. "It is against the objective rendering of the emotions—this thrusting into the region of fact and positive knowledge of conceptions essentially ideal and poetic—that science . . . wages war."[27] And here precisely is where evolutionism, including its social Darwinian element, failed. Indeed, it did not even attempt to compete, because it had other goals: to promote a world picture, an ideology of progress.

An ideology, to be sure. But would the term "religion" also be appropriate? Considering the nature of the beast, it truly seems so. The concept of a religion is notoriously hard to define, but one thinks in terms of a world picture, providing origins, a place (probably a special place) for humans, a guide to action, a meaning to life. There are other prominent features of many religions, such as belief in a deity and a formalized and recognized priesthood, but these features are not absolutely essential to the definition. Buddhists (and many Unitarians) would probably flunk the God question, and Quakers (by explicit design) have no clergy. Rather than getting too flustered by counterexamples, let us allow the oxymoron "secular religion" and cast our question in these terms. And the answer does seem positive.

Popular evolution—evolutionism—offered a world picture, a story of origins, and a special place for humans in the scheme of things. At the same time, it delivered moral exhortations, prescribing what we ought to do if we want things to continue well (or to be redeemed and a decline reversed). These things hardly came by chance or in isolation. In asking about origins, evolutionism was answering a question posed by Christianity (and Judaism before this), and in focusing on the status and obligations of humans, evolutionism was trying deliberately to do better than Christianity.

The followers of Auguste Comte founded a "religion of humanity" based on positivism, with all sorts of silly practices like the naming of secular saints. Evolutionists, to their credit, did not go this far. In fact, they took much care not to go this far, for formal groups of nonbelievers (secularists) tended to have the stench of working-class radicalism and nonrespectability that scientists and their supporters—striving hard to make themselves indispensable (and powerful) members of Victorian society—were at great pains to avoid. Huxley disapproved personally and politically of Charles Bradlaugh, who, although elected a member of Parliament, was unable for many years to assume his seat because of his refusal to take a Christian oath of allegiance. Herbert Spencer refused to cross the Atlantic on the same ship as George Jacob Holyoake, a notorious freethinker: "I think it very probable that the fact of our association would be used against me."[28]

Self-serving though this kind of behavior may have been, it fulfilled its end. In his novel *Born in Exile*, George Gissing wrote sardonically of "that growing body of people who, for whatever reason, tend to agnosticism but desire to be convinced that agnosticism is respectable; they are eager for antidogmatic books to be written by men of mark. They couldn't endure to be classed with Bradlaugh, but they rank themselves confidently with Darwin and Huxley."[29] Yet, for all that they wished to remain respectable, the evolutionists still could not forgo the sweet taste of morality. And so they settled for a religion substitute, or religion lite. Not only did Huxley sit comfortably at the top of the hierarchy of a "priesthood of science"—the press referred to him as "Pope Huxley"—but his letter to Kingsley shows that, for him, hands-on contact with the world substituted for Holy Communion. Darwin, particularly after being buried in Westminster Abbey, came pretty close to divine status. (History does not record whether disciples were hanging around three days later.)

Strengthening the case for the religious nature of post-Darwinian evolutionism was its link to millennial thinking, which began to flourish in a major way in popular culture in the nineteenth century.[30] Many Protestants took the French Revolution and the Napoleonic wars to be signs of great significance, especially given their negative implications for the Catholic Church. Perhaps the Antichrist (that is, the Pope) was now truly threatened. Those drawn to postmillennial varieties of apocalyptic prophecy, in which Christ comes after the millennium, generally had optimistic views of human nature. The brighter future was something to be achieved by us, rather than something imposed from above. We have seen that this kind of thinking was akin to, if not outrightly identical to, progressivism of one sort or another. As in the century before, Christian eschatology and progressivist philosophies were often closely connected. Remember the tripartite division of Joachim of Flora and its echo in the theological/metaphysical/scientific schema of Comte.

Because the Second Coming was now postponed to some time in the indefinite future, increasingly such pictures appealed to people with few or no religious commitments. In Britain, the man who so influenced Alfred Russel Wallace, Robert Owen—who rejected all formal religious systems—was greatly influenced by millennial ideas. He famously (or notoriously) tried to set up a planned utopia in the New World.[31] At times, Owen took on the mantle of the Messiah himself, writing of being compelled "to proceed onward to complete a mission" whereby "the earth will gradually be made a fit abode for superior men and women, under a New Dispensation, which will make the earth a paradise and its inhabitants angels."[32]

Even those who did not have such pretensions joined in the general theme. The Victorians loved to gather around the piano for a good sing.

> There's a good time coming, boys,
>   A good time coming:
> There's a good time coming, boys,
>   Wait a little longer.
> We may not live to see the day,
> But earth shall glisten in the ray
>   Of the good times coming.
> Cannon-balls may aid the truth,
>   But thought's a weapon stronger;
> We'll win the battle by its aid:
>   Wait a little longer.

This kind of sentiment continued unabated into the post-*Origin* part of the century, except that now such thinking was fused with evolutionism. Typically, Herbert Spencer denied that his thinking was influenced by predecessors—he pretended that he reviewed a book by Comte that he had not read because he found the contents so distasteful. But even if Spencer was telling the truth, others were in the stream of history. Wallace was not the only one who breathed a kind of inherited, secular postmillennialism (for many, as we shall see, not so very secular), with human effort being urged to bring about a brighter future. A form of evolution-based secular millennialism, complete with religious language and metaphors, was frequently invoked. Although identifying Darwin's bulldog (as Huxley was known) with the pontiff was intended to capture the way in which Huxley led the movement of science, one suspects that one or two in their country parsonages did think of Huxley as the Antichrist.

Admittedly, all sorts of different and contradictory things were being claimed in the name of progressive evolution, but in this respect it was no different from regular religions. Some Christians defend war on the basis of the gospel, others argue for pacifism; some defend capitalism, others argue for socialism; and some cling to female subordination (as with an all-male priesthood), whereas others believe that Christianity mandates gender equality. The point is that—for the secular millennialist—one must do something to bring about change and improvement.

Looking at matters another way, in Britain particularly many people, including professional scientists, wanted nothing at all to do with formal religion, and they saw science and religion locked in a deadly battle. As Huxley put it back in 1859, "Theology and Parsondom . . . are in my mind the natural and irreconcilable enemies of Science. Few see it but I believe we are on the Eve of a new Reformation and if I have a wish to live 30 yrs, it is to see the God of Science on the necks of her enemies."[33] The years after the *Origin* saw such bestsellers as *The History of the Warfare between Science and Theology in Christendom.* Although the trial of Galileo would always be the main course in such orgiastic feasts as these, Darwin provided a tasty dessert—at least Darwinism did. Huxley wrote that new science has to demolish theological obstructions, rather as the infant Hercules strangled snakes while in his crib. For people subscribing to that nonbelief newly coined as agnosticism, evolutionism—with its progress, origins, human status, and exhortations—was just what was wanted in the new age.

Given his views on the Unknowable, Herbert Spencer could hardly be described as a materialist, but many evolutionists could. They were all at one in wanting to take God out of the equation and to let the laws of nature govern everything. Knowing that he had nothing of value to say on the origin of life issue, and knowing also (he wrote just at the time when Pasteur was putting the final boot into spontaneous generation) that critics would pounce on ignorance and wild speculation, Darwin wisely had said nothing on the subject in the *Origin.* But this did not stop others from jumping into the fray. In a way, they had to, because materialism was part of their overall picture. They were not just offering a theory, they were offering a system—a system with a beginning and a hoped-for end, an eschatology, no less than the system they were trying to replace.

By the last decade of his life, Thomas Huxley—who had deliberately called a set of his essays "lay sermons"—was losing faith in progress. But his ardor to see evolutionism as the proper alternative to the Christian religion burned no less brightly.[34] Thirty years had now passed since he had spoken of bringing on a new Reformation, and he was not now about to abandon the God of Science. But at this late point, a truly curious episode in Huxley's controversial life occurred. He attacked the Salvation Army and its leader, "General" William

Booth, with a ferocity reminiscent of his long-ago attack on Chambers's *Vestiges*. In lengthy letters to the *Times*, Huxley accused Booth and the Army of just about every sin in the book—totalitarianism, religious bigotry, fiscal mismanagement, and more. "Undoubtedly, harlotry and intemperance are sore evils, and starvation is hard to bear or even to know of; but the prostitution of the mind, the soddening of the conscience, the dwarfing of manhood are worse calamities."[35]

Huxley wrote not one word about the truly saint-making work that Booth and his followers had been doing with the poor and despised for over twenty years. Not one word about the soup kitchens and the beds, the care of drunkards and the sick and the truly desperate in society. Why? A little book collecting the letters, published by Huxley with the provocative title *Social Diseases and Worse Remedies*, gave away the game. The introduction was a reprinted essay about the virtues of science, and science education, and of moving from bigotry and ignorance, especially Christian bigotry and ignorance. Huxley (rightly) saw the Salvation Army as competing for his space. It was not part of the establishment—anything but. Like Huxley and his fellow evolutionists, the Army was trying to address the failings that had come from the inadequacy of the establishment. Yet the Army was doing it in exactly the opposite way—by sermons and sympathy, meetings and songs and brass bands and free grub rather than by workingmen's lectures and education; by fervent reliance on a simple reading of the Bible rather than by observation, experiment, and reason; by enrolling as its officers (its clergy) people from the lower ranks and impressing on them the need for conformity and obedience and a simple heart, rather than years of laborious, intellectual training. And what was worse, as Huxley's appalled reaction showed, the Salvation Army was having success and gaining credibility—especially with sympathizers among liberal Christians who were friends of the scientists and with the lower masses to whom Huxley's scientific priesthood was likewise catering.

To use a phrase invented by Thomas Henry Huxley's biologist grandson, Julian Huxley, the evolutionists were truly in the business of providing a "religion without revelation"—and like all fanatics, they were intolerant of rivals.

7

Christian Responses

A fire-mist and a planet
A crystal and a cell
A jelly-fish and a saurian
And caves where the cave-men dwell;
Then a sense of law and beauty
And a face turned from the clod –
Some call it Evolution,
And others call it God.

*T*his poem, "Each in His Own Tongue," by the turn-of-the-nine-teenth-century Kansas poet William Herbert Carruth would surely make anyone's list of the world's ten worst poems. One would like to be able to speak of Carruth as "justly neglected," but unfortunately that was not the case. His verses have long been a staple of devotional anthologies, favored graduation gifts at Christian high schools. (I speak with some experience in this matter.) Before we move to the ideas at the heart of the Christian response to evolutionism, let us pause to marvel at one last Carruthian verse—these lines from the sonnet "Weeds."

Patient the long day
Ye take the buffetings of scornful clay,
Sustained by that small portion of God's dew
Which thick-strewn dust permits to fall on you.

And live where finer herbs must wilt away.
Have ye too, dreams of better things to be?
Of worlds in which the crooked shall be straight,
Where all that are in bondage shall be free,
And lifted up all thowse of low estate?

Before Charles Darwin, hardly anyone really wanted to take God out of the discussion of design in the natural world, and those that did—Diderot, for example—really had no right to do so because they had no substitute to put in God's place. After Darwin came along and spoke about final cause, what then happened to God? Were evolution and God not locked in mortal combat? Apparently Carruth's weeds were Christian Providentialists who dreamed of the day when, through God's grace, they would be raised up to the status of flowers. Then they too would make the cut and be arrayed in the Divine Vase. But what of those weeds that were cultural progressionists, who dreamed of the day when, through their own efforts, they would rise up to the status of flowers? Is there hope for them, too? Or are they doomed to wither and fade, rejected by that great process governing us all?

> Some call it Consecration,
> And others call it God.

Bishop Wilberforce was not the only opponent of evolution. Whewell acknowledged politely his copy of the *Origin* (all of the old guard were sent complimentary copies) but opposed it in the preface to a late edition of his Bridgewater Treatise and supposedly would not allow the *Origin* on the shelves of the library in Trinity College. Sedgwick went to his grave in ardent opposition, having told Darwin that he laughed, he cried, he disagreed. As far as he was concerned, the

traditional arguments of natural theology—that design in the organic
world argued for a designer—continued to rule and were taken as a
direct counter to Darwinism. His "deep aversion to the theory" was in
major part based on the fact that it "utterly repudiates final causes."[1]
Many good English establishment Christians were appalled at what
they rightly saw as a threat to a Bible-based religion, although the dis-
mantling of the creation stories of Genesis was not their major concern. The real upset was rather the pressure on the more Jesus-cen-
tered aspects of Christianity—the incarnation, the resurrection, and
above all the promise of eternity, either eternal life or eternal punishment.

Today, it seems a little odd that eternal damnation would have had
such theological significance. In his *Autobiography* Darwin made clear
that this doctrine was what he most disliked about Christianity. Nevertheless, it was the really big issue among Englishmen. It had long
been a major part of Calvinist theology. In the eighteenth century,
Jonathan Edwards's best-known sermon, "Sinners in the Hands of an
Angry God," had left little to the imagination. "The God that holds
you over the pit of hell, much as one holds a spider, or some loath-
some insect over the fire, abhors you, and is dreadfully provoked: his
wrath towards you burns like fire; he looks upon you as worthy of
nothing else, but to be cast into the fire; he is of purer eyes than to
bear to have you in his sight; you are ten thousand times more abomi-
nable in his eyes, than the most hateful venomous serpent is in ours."
But the persistence of the threat of eternal damnation down through
the nineteenth century, in England no less, seems quite odd.

One is tempted to offer a psychoanalytic explanation for its importance. Most upper-middle-class Englishmen had been to private
schools (misleadingly known as public schools), where corporal punishment was all-pervasive; and (experience speaks again) they tended
to regard God as a headmaster (many of the bishops had been head-

masters). The major reason, however, is that in the nineteenth century, as in the eighteenth, God's justice was bound up with the social role that Christianity was thought to play. Eternal punishment was needed to reinforce morality in Britain and (increasingly) in its growing empire. Even if the governed were not Christian, the administrators had to be. They needed a moral foundation to sustain themselves and to rule the natives: "Ronny's religion [referring to the pathetic magistrate in *A Passage to India*] was of the sterilized Public School brand, which never goes bad, even in the tropics."

Yet, as our earlier survey of religion before the *Origin* led us to expect, despite opposition to evolution from many practitioners of both revealed and natural theology, the initial reaction by Christians was far from uniformly negative. The Reverend Baden Powell immediately endorsed the message of the *Origin*: "Mr Darwin's masterly volume on *The Origin of Species* by the law of 'natural selection,'—which now substantiates on undeniable grounds the very principle so long denounced by the first naturalists,—*the origination of species by natural causes:* a work which must soon bring about an entire revolution of opinion in favour of the grand principle of the self-evolving powers of nature."[2]

There were others. Already mentioned was the Harvard botanist Asa Gray, who may not have bought into the whole message but went a very long way. Because Gray wanted to supplement natural selection with some kind of guided variation, Darwin thought that he was selling out—which was true—but Gray's commitment to evolution of a kind was solid. And in fact, something along the lines of Gray's compromise grew increasingly attractive to many Christians. They obviously could not buy into Huxley's agnostic materialism, and Spencer's Unknowable was not a great deal better. But they did want to be on the side of modern science, and in some respects evolution was attractive. It solved a number of scientific problems, and by speaking of origins it did take seriously the sorts of questions that Christians thought should be taken seriously.

Entirely typical was the Duke of Argyll, head of the Clan Campbell, patrician politician, father-in-law of one of Victoria's offspring, general man of science. He had little time for unaided natural selection but granted a sort of guided evolution. "Creation by Law—Evolution by Law—Development by Law, or, as including all those kindred ideas, the Reign of Law, is nothing but the reign of Creative Force directed by Creative Knowledge, worked under the control of Creative Power, and in fulfilment of Creative Purpose." Other Christians facing evolution—often people of a rather traditionalist disposition—went even further, arguing that evolution is a good thing for the Christian rather than a hindrance. Later in the nineteenth century, the Anglo-Catholic Aubrey Moore claimed that nothing in Darwin's thinking threatened faith, and indeed the theory brought God back into the world rather than the converse.[3]

Likewise, the American conservative Presbyterian (and friend of Asa Gray) George Frederick Wright, who wrote: "The student of natural history who falls into the modern habits of speculation upon his favorite subject may safely leave Calvinistic theologians to defend his religious faith. All the philosophical difficulties which he will ever encounter, and a great many more, have already been bravely met in the region of speculative theology." He added: "The doctrine of the continuity of nature is not new to the theologian. The modern man of science, in extending his conception of the reign of law, is but illustrating the fundamental principle of Calvinism."[4]

The New Natural Theology

Helping Christians take this stand was the fact that virtually no one took natural selection seriously as the sole (or major) causal factor in evolutionary change. The scientists of the day offered Lamarck-ism, saltationism, orthogenesis, and much more as alternatives, and wrapped all of this up with a progressivist bow. Although tradition-ally progress was thought to oppose Providence, increasingly Chris-tian Victorians showed a willingness to flow with the tide. Without denying that the salvation of human souls required miraculous inter-vention, one could substitute design through evolution (progressive evolution, that is) for Providential design through adaptation. The move was from a God-inspired, short-term, adaptive teleology to a God-inspired, long-term, historical teleology: from the planned ends of organisms to the planned end of creation.

Frederick Temple, future archbishop of Canterbury, writing in the 1880s, was quite explicit on the need to make this shift from an old-style, natural theology to a new-style, evolution-informed substitute. He stressed that, after Darwin, if one were rewriting Paley's *Natural Theology*, the emphasis would be different. "Instead of insisting wholly or mainly on the wonderful adaptation of means to ends in the struc-ture of living animals and plants, we should look rather to the original properties impressed on matter from the beginning, and on the bene-ficent consequences that have flowed from those properties."[5]

As in the science itself, often it is difficult to tell if someone has made the move to evolution.

> God renews
> His ancient rapture! Thus He dwells in all,
> From life's minute beginnings, up at last
> To man—the consummation of this scheme
> Of being, the completion of this sphere
> Of life: whose attributes had here and there
> Been scattered o'er the visible world before,
> Asking to be combined, dim fragments meant

To be united in some wondrous whole,
Imperfect qualities throughout creation,
Suggesting some one creature yet to make,
Some point where all those scattered rays should meet
Convergent in the faculties of man . . .

Thus the young poet Robert Browning in 1835, in an oft-quoted poem. Is this evolutionary or not? Does it matter?

Across the Atlantic, the new president of Princeton was James McCosh, a Presbyterian divine who began in Edinburgh, moved to Northern Ireland, and then went to the New World. Sensitive to science and yet deeply committed to his faith, he too saw evolution as the key to a new natural theology, not just the killer of the old. He invited his readers to "look upon evolution as we do upon gravitation as a beneficent ordinance of God. Gravitation is a law of contemporaneous nature, binding the bodies in space. Evolution is a law of successive nature, binding events in time. The two are powerful instruments in giving a unity and consistency to the world, and in making it a system compacted and harmonious, admired by the contemplative intellect."[6]

He spelled out in detail the connection between progress, evolution, and design. "There is proof of Plan in the Organic Unity and Growth of the World. As there is evidence of purpose, not only in every organ of the plant, but in the whole plant . . . so there are proofs of design, not merely in the individual plant and individual animal, but in the whole structure of the Cosmos and in the manner in which it makes progress from age to age." But do not think that any of this happens by chance. God had to arrange and drive everything, including—especially including—those things covered by law. "All such laws are complex . . . [but] the law of the progression of all plants and of all animals is still a more complex one, implying adjustment upon adjustment of all the elements and all the powers of nature towards the accomplishment of an evidently contemplated end, in which are displayed the highest wisdom and the most considerate goodness."[7] A similar philosophy was pushed at the popular level. The Reverend Henry Ward Beecher, a charismatic preacher and brother of the novelist Harriet Beecher Stowe, took progress as his theme, if not his obsession. "If single acts would evince design, how much more a vast universe, that by inherent laws gradually builded itself, and then created its own plants and animals, a universe so adjusted that it left by the way the poorest things, and steadily wrought toward more complex, ingenious, and beautiful results!" Continuing: "Who designed this mighty machine, created matter, gave to it its laws, and impressed upon it that tendency which has brought forth the almost infinite results on the globe, and wrought them into a perfect system? Design by wholesale is grander than design by retail."[8]

Beecher knew first-hand about sin. He was one of the most notorious adulterers of the nineteenth century. But in his faith Beecher was in no way half-hearted or insincere or in doubt that Jesus had died for his transgressions. Beecher was a genuine Christian. In part precisely because they were Christians, Beecher and the others were striving to mold their beliefs to fit the modern world. They wanted to be on the side of science, and that included evolutionary science. In classic postmillennialist fashion, they believed that God expected people to do science and use it for the good of mankind: "In all the geological ages we find in any age the anticipation of the following. This may also be the case with the age in which we now live, the Age of Man. We see everywhere preparations made for further progress, seeds sown which have not yet sprung up; embryos not yet developed; life which has not yet grown to maturity. In particular, we find that in this Age of Man, man has not yet completed his work."[9]

McCosh, as reforming president at Princeton, made major and

very successful efforts to upgrade science teaching at his institution. Several of his students were packed off to study with Huxley no less, and why not? Darwin's bulldog was the best teacher in the business.

Revitalizing Christianity

Science was not the only factor in the equation. Independently, people were searching for ways to revitalize or even replace traditional Christianity. Socially and economically, as Britain grew from a small, essentially rural society to a much larger, urban society, the religious supports and justifications of social policy started to fray and tear apart. Christians, particularly evangelicals like Thomas Chalmers, seized with enthusiasm on Malthusian claims. They saw the hand of God at work here, forcing people to labor and make an effort to earn their daily bread, and at the same time pointing to sexual restraint as the right way to avoid poverty. That Malthus also pointed away from progress, for all human-produced gains are bound ultimately to come to naught, only reinforced the truth of his conclusions.

As part of their general thinking, Chalmers and other evangelicals argued strenuously against state support for the poor, for this would only exacerbate the problem. Personal charity—and to be fair, many evangelicals gave huge amounts of money and time to helping the needy—was the only right way forward. While this politico-theology may have worked more or less in 1800, by 1850 its limitations were there for all to see. In the mid-1830s, draconian new Poor Laws had been enacted to get the poor to work, or else. Nassau Senior, a political economist, put it bluntly to Henry Brougham, politician and natural theologian: "No relief shall be given to the able bodied, or to their families, except in return for work, and that work as hard as it can be made, or in the workhouse, and that workhouse as disagreeable as it can be made."[10] As *Oliver Twist* shows well, those workhouses could be made very disagreeable indeed, and in *Our Mutual Friend* the desperate efforts of Betty Higden not to fall upon (the mercies of) the parish show that, at the individual level, the laws were successful in keeping people from state relief.

Unfortunately, as the Irish famines of the 1840s brutally proved, private charity simply could not handle the major disasters of a nation with such a large population. The state had to get involved in providing relief, and hence the old theological underpinnings needed revision and more. Not revision of the Malthusian calculations, obviously—these inferences were more pertinent than ever—but revision of the assumption that the consequences reflected God's good intentions. To their credit, evangelicals themselves realized this. Increasingly, they saw it as wrong to blame all of the troubles of the Irish on their being Catholic and thus deserving of God's punishment. Eating potatoes rather than corn is no heresy. Moreover, these Christians were acutely aware of the obscene irony that the easiest and most efficient way of raising large funds for the suffering usually involved much jollification, as at charity balls. Surely it was not part of God's plan that one group of citizens should gorge themselves on meat and drink because another group was starving.

Intellectual threats only exacerbated the social tensions that were undermining traditional religious thought. Biblical criticism was now running at full throttle in the English-speaking world. In 1861 in Britain, a group of more advanced-thinking Anglicans (seven, of whom six were ordained, including Baden Powell, Temple, and Benjamin Jowett, the classicist and future master of Balliol College) published a volume, *Essays and Reviews,* which made full and favorable use of Germanic findings about the history of the Bible.[11] Casting doubt on miracles and the like, this kind of work in itself made Christianity more science-friendly.

Science benefited in other more coincidental ways. In the huge controversy that followed the volume's appearance, some worried that this detracted from the Darwin debate. "The book [the *Origin*] I have no doubt would be the subject still of a great row, if there were not a much greater row going on about *Essays and Reviews*."[12] Two of the essayists were prosecuted for heresy for having denied the Church teaching of eternal punishment. When the defendants were found not guilty, a wit declared that the gravestone epitaph of the judge would read:

> He dismissed Hell with costs,
> And took away from orthodox members of the Church
> of England
> Their last hope of everlasting damnation.

As the century drew on, theologically as well as legally, notions like substitutionary atonement and eternal punishment came increasingly under attack. The emphasis shifted from Easter to Christmas, from Christ as sacrificial lamb to the infant Jesus as the embodiment of peace on earth, good will toward men—from atonement to incarnation. It seemed ethically distasteful that God should have to suffer for our sins, and contrary to his goodness that even the worst of sinners should suffer forever. As ideas like these declined in significance, so did the urge to tie faith tightly to a holy text.

Matthew Arnold was among those searching for a way to bring Christianity into the new age.[13] As a student at Oxford, he had been deeply struck by John Henry Newman—not so much the content of his message as the style in which it was delivered, in contrast to evangelical emotion. Although Newman felt decidedly ambiguous about the debt—with wry humor, overlaying serious intent, he told Arnold that he was praying for his conversion to Catholicism—the older man's influence reinforced Arnold's strong aversion to any kind of enthusiasm in religion. For thirty-five years Arnold was a government inspector of nonconformist schools, and he grew to hate the narrow-minded bigotry that so often accompanied dissent.

Arnold made a distinction between Hebraism and Hellenism, the former being marked by "strictness of conscience" and the latter by "spontaneity of consciousness." He had been brought up on a diet of the classics; his father, Dr. Thomas Arnold, was the greatest force in British secondary education in the nineteenth century. As Oxford's professor of poetry, Matthew had lectured on Homer and would re-lax at the end of a hard day's work with a couple of hundred lines of Greek poetry in the original. There could be little doubt about where Matthew Arnold's preferences lay. Telling the sad tale of a Mr. Smith who committed suicide because he feared poverty and eternal damna-tion, Arnold berated the Philistine nature of a religion that reduces everything to making money and saving souls—a theology he saw as too prevalent in evangelical circles.

Arnold was a close friend of Thomas Huxley—the two were both very much in the state-education business—and he agreed with the scientist's insistence on the virtues of Bible reading for moral instruc-tion. In taking that position, Huxley certainly did no harm to the evo-lutionists' quest for social respectability. He wrote that he had been "seriously perplexed to know by what practical measures the religious feeling, which is the essential basis of conduct, was to be kept up, in the present utterly chaotic state of opinion on these matters, without the use of the Bible." Arnold never followed Huxley into agnosticism. He was a churchgoer, writing: "Two things about the Christian reli-gion must surely be clear to anybody with eyes in his head. One is, that men cannot do without it; the other is that they cannot do with it as it is." All of the stuff about miracles was simply no longer tenable. But we must have some basis of ethical discourse. "In morals, which are at least three fourths of life," to do without the Bible and Christianity is "exactly like doing in aesthetics without the art of Greece." Arnold himself ended up close to a kind of Spinozistic pantheism, where "our God is the eternal not ourselves that makes for righteousness," about whom we may speak in biblical terms but "approximative and poetical merely."[14]

The pressure was to find a demythologized Christianity that spoke to the social and ethical issues of the day. For many, the new science of evolution was literally God-given. Even across the Atlantic, liberal thinking was moving away from traditional theological claims and in the direction of ethical philosophy. By century's end, influential church figures preferred to talk in terms of a "social gospel" than of original sin and atonement and everlasting life. "Religious morality is the only thing God cares about." Influenced by secular notions of progress, Christians wanted to create—and thought they could create—God's kingdom here on earth. We must therefore strive toward a "Christianizing of the social order."[15]

The generation following McCosh and Beecher continued in the postmillennialist tradition. These liberal Christians (no less than the secular thinkers) scorned literalist readings of Daniel and Revelation, or any other books of the Bible, for that matter. They took seriously the findings of higher criticism, which in turn led to highly allegorical interpretation. But in major respects, the Christians, no less than secular thinkers, embraced postmillennialism as the philosophy of "modernists." They saw a good time coming, so long as we make the effort. They saw Jesus "coming in cleaner politics, better industrial conditions, purer recreations, a fairer distribution of wealth, a more wholesome social life, a greater kindliness and kindly consideration for each other, in the abolition of vice, in a permanent and richer peace, and best of all in a deeper, richer, growing religious consciousness."[16] A good time coming that linked naturally with enthusiasm for a progressivist evolution—the progress of biology supporting and reflect-

ing the progress of society, and conversely. A Christianized evolutionism, or an evolutionized Christianity.

Roman Catholicism

Meanwhile, what about the Catholics? To put it bluntly, evolution was
not their fight.[17] Their fingers had been burned two centuries before,
over that mess with Galileo, and they were glad to let the Protestants
run with this one. But there were also social reasons why evolution was
not in the Catholics' sights, except in France, where Darwinism was
pushed as part of a materialist-positivist program. In nineteenth-century Britain and in America, Catholics were generally at the bottom of
the social structure—the Irish in Britain, and the Italian, Polish, and
Irish immigrants in America. Their concern was to make a living and
raise their family in an alien culture, not to engage in esoteric debates
about science and religion.

But with these qualifications, at one level Catholics took the same
attitude toward evolution as did Protestants. Positivism, naturalism,
and materialism were totally unacceptable—one could not imagine
that immortal souls were things that just happened. There had to be
divine intervention there. But beyond that, Catholics went with science, or at least the popular science that evolution had become. This
was the position of John Henry Newman. We have seen that he never
had much taste for natural theology, so that was no barrier. The same
was true of the literal words of the Bible: "The Fathers are not unani-
mous in their interpretation of the 1st chapter of Genesis. A com-
mentator then does not impute untruth or error to Scripture, though
he denies the fact of creation or formation of the world in six days, or
in six periods. He has the right to say that the chapter is a symbolical
representation, for so St. Augustine seems to consider."[18]

On the positive side, when Newman was working to set up a Catholic university in Dublin in the 1850s, he spoke strongly of the need to keep scientific inquiry unfettered by religious dogma. His own position was that science and religion deal with different spheres and thus, properly understood, do not interact and cannot conflict. Speaking of the natural and supernatural worlds, he said that "it will be found that, on the whole, the two worlds and the two kinds of knowledge respectively are separated off from each other; and that, therefore, as being separate, they cannot on the whole contradict each other."[19]

Significant for Newman's attitude toward evolution was his developmental attitude toward knowledge and belief. In the world of ideas, of religious ideas, he did not want change to negate earlier truths, but he found no problems at all with biological evolution per se. To the conservative Anglican Edward Pusey, in support of Darwin's receiving an honorary degree from Oxford University, Newman mused: "Is this [Darwin's theory] against the distinct teaching of the inspired text? if it is, then he advocates an Antichristian theory. For myself, speaking under correction, I don't see that it does—contradict it."[20] Newman was not enthused by the details of Darwin's theory, and he fell in with the general opinion that there had to be more to it all than unaided natural selection. Newman was certainly no easy cultural progressionist, but in the world of life he wanted firm direction toward human beings.

Given this kind of thinking, Newman was naturally sympathetic to the science of the Englishman St. George Mivart—a Catholic convert and sometime student and disciple of Huxley. Mivart's *Genesis of Species*, published in 1871, argued with much energy that any theory based on natural selection was riddled with problems. As an alternative, Mivart—who never challenged evolution as such—supposed (with Huxley) that there were occasional major jumps or leaps, called saltations, and that (contrary to Huxley) these jumps are directed toward adaptive advantage—adaptive advantage that is the design product of a good god.

Souls, of course, are created miraculously. But for the rest, Mivart's picture is one of God-driven upward thrusting toward our own species, *Homo sapiens*. As he wrote later in the decade: "The process of evolution, as carried through the material world, shows us the evolution from potentiality into actuality of successively new forms. We cannot imagine how they are produced; we simply recognise that they are. In passing to the vegetable world from the mineral kingdom, we behold, for the first time manifested, a sentient form. In passing to the human world from the kingdom of brute animals, we behold, for the first time manifested, a rational form." And let there be no mistake, evolution or not, we are into something altogether new and better. "With our entrance on the world of self-consciousness, reason and free volition, we impinge upon another order of being from that revealed to us by all below it—an order of being which the cosmical universe, as it were, intersects, as the different lines of cleavage and stratification may intersect in the same rock."[21]

The rest of the nineteenth century saw the Catholic position move backward. By the beginning of the twentieth century evolution was proscribed, and John A. Zahm, a science professor at Notre Dame, among others, was forbidden to teach evolution. Mivart, who had at first been honored by the Vatican, died excommunicated, although as much for his propensity to quarrel with everyone as for his science. The issue for Catholic authorities was not evolution per se but a more general move away from the modern world. This came about in part as a function of the unification of Italy, with its corresponding loss of the power (and land) of the Holy See, and the latter's retreat into a more rigid and super-traditionalist mode.

The Syllabus of Errors of 1864 denied that "the Roman Pontiff can and should reconcile himself to and agree with progress, liberalism and modern civilization." Then, at the beginning of the 1870s, the Church decreed that, on matters of dogma, the pope speaks infallibly. What followed from the Vatican was an exercise of power over the far-flung church, particularly in the New World, where bishops had become accustomed to considerable autonomy. Thomas Aquinas was declared the official touchstone for Catholic theology and philosophy—somewhat of an irony, since in his own day Thomas had run into trouble for his avant-garde thinking. Alternative systems were rejected. These included anything that might threaten the Vatican's vaunted authority, and thus evolution (the epitome of "progress, liberalism and modern civilization") was downplayed and denied. The proscription started to lift only in the middle of the twentieth century.

Today, Mivart's theological line on evolution would find strong support in the Vatican. Although Pope John Paul II has insisted that the arrival of the immortal soul demands a miracle, he has been explicit in his endorsement of an evolutionary view of nature. Catholicism has embraced even Darwin, but it was a long time coming.[22]

8

Fundamentalism

Gimme me that old time religion
Gimme me that old time religion
Gimme me that old time religion
It's good enough for me.

It was good for little David
It was good for little David
It was good for little David
And it's good enough for me.

It was good for the Hebrew children
It was good for the Hebrew children
It was good for the Hebrew children
And it's good enough for me.

*T*o the strains of this American folk hymn (which has many additional verses), the authorities of the town of Hillsboro—the mayor, the preacher, the court officers—gathered and, with a photographer to record events, walk across the square to the schoolroom where Bertram T. Cates is teaching his biology class—the class where he tells his students that we are all descended from a lower order of animals. Arrested for breaking the law that bans the teaching of evolution, Cates becomes the center of a trial that attracts the attention of the world. He is prosecuted by three-times presidential candidate Matthew Harrison Brady, defended by the noted agnostic lawyer

Henry Drummond, and reported on by cynical newspaperman E. K. Hornbeck.

The scene is fictional, taken from the 1960 film version of the play *Inherit the Wind*, with Frederick March playing Brady, Spencer Tracy playing Drummond, and Gene Kelly playing Hornbeck and hogging the best lines. (To Cates's girlfriend, who objects to his joking about serious matters: "I may be rancid butter, but I'm on your side of the bread.") At another level, we are far from fiction, for the play is based on one of the most notorious trials of the twentieth century, the so-called Scopes Monkey Trial, during which a young schoolteacher was prosecuted for teaching evolution by a three-times presidential candidate, defended by a noted agnostic lawyer, and reported on by a cynical newspaperman.

If so many after the *Origin* were being so accommodating to evolution—suggesting, which is certainly true in part, that the supposed war between science and religion was a fiction promoted by the evolutionists to make their case seem more compelling and urgent—then how did this trial come about? Even more pressing, why do we have creationists in America still today? To answer these questions, we have to pick up on American religion around 1800.

When we left things, there were factors and reasons suggesting that religion was going to be one of those Old World phenomena—like monarchy—that the new nation was casting aside. For two centuries, America had been intensely religious, a propensity that had surged in the mid-eighteenth century with the Great Awakening. But the key figures in the Revolution had been nominal Christians, if that. They were drawn to deism (or less), thinking that religion was a private matter with a minimal state-endorsed, public role. Religion in Europe was already into its long decline. Why would not the same be true, even more so, in America?

The very opposite happened.[1] By the beginning of the nineteenth century, the United States was set for explosive religious growth. Not so much in the old sects like Presbyterianism and Congregationalism and Quakerism, and certainly not in non-Protestant denominations such as Catholics and Jews, who would arrive in great numbers later in the century, but in the strongly evangelical sects, the Baptists and, above all, the Methodists, that branch of Christianity which responded to the Enlightenment by turning away from reason toward faith and emotion.

In 1790 there were 712 Methodist churches in the United States; by 1860 the number had grown to 19,883. Baptist churches increased from 858 in 1790 to 12,150 in 1860. Presbyterians and Congregationalists (the old Calvinist churches) went up too, but by nothing like the same rate: Presbyterians, from 725 in 1790 to 6,406 in 1860; Congregationalists, from 750 in 1790 to 2,234 in 1860. Looking at things another way, in 1790 the population of the United States was about 4 million and the number of Methodists around 60,000 (1.5 percent of the population). By 1860 the population of the United States was over 30 million and the number of Methodists was almost 2 million (5.5 percent of the population). If splinter groups and black churches are included, then the figures are significantly higher.

The big growth occurred in the South, the mid-West, and the far West, not in New England. In the South, for instance, almost 40 percent of the Christian population was Methodist, with about that many Baptists. Episcopalians made up a mere 5 percent. By 1860 about a third of the nation's citizens were actively connected with churches, nearly all of which were Protestant and 85 percent of which were evangelical. The seating capacity of these churches suggests that as much as another third of the population attended church occasionally. The contrast with trends in the Old World is stark. As early as the 1830s, church attendance in America was proportionately twice that in Britain.

What explains this Second Great Awakening in the nineteenth century? The key has to be the Revolution and its aftermath. The Revolution broke the power of the established churches and created ecological niches for those with the energy and interest to occupy them. More than this, it made unpopular any church with a hierarchy and with authority exercised from above. This leveling of the playing field gave an opportunity to religions that emphasized individual commitment, regular Bible reading, and a personal relationship with God. Methodism particularly, with its belief that the individual's willingness to step forward and accept Christ's sacrifice on faith was crucial to salvation (as opposed to stricter Calvinist doctrines emphasizing predestination), was very attractive in a new nation built on ideals of individualism, free will, and egalitarianism.

Central to this surge of religious feeling was the almost universal belief that America was not just another country. It was, rather, God's chosen land, and Americans were God's chosen people—the new Israel. The Mormons were just an extreme version of this exceptionalist vision. Americans had a new covenant with God. Yet, exciting though it may be to build a new Jerusalem, it would take a lot of hard work; Americans faced a dangerous slog, virtually on a daily basis, particularly in the South and in the expanding West. Lacking the established rules and comforts of tradition—a fixed social order, a religion with the legal and moral force of the state, the folkways and myths of the village—the young nation had to put in its place something that would make the effort worthwhile, and at the same time help to regulate both public and personal life on the frontier. The Protestant religion—usually promoted by charismatic and dedicated preachers—moved right in to perform all of this crucial social work. Literally,

Protestantism, with its admirable work ethic coupled to its vision of faith, helped to build the nation.

But what exactly was that vision of faith, among Methodists especially? How was one's belief in Christ put into action? How was one supposed to act, not so much to buy salvation but to show that one had been saved? How should fathers relate to their wives and families, and what were the duties and expectations of mothers? What governed disputes between neighbors, or between bosses and workers? Protestants had always looked to the Bible for guidance, and Calvinists particularly had let their actions be ruled by what they found there. Not that Calvin or his followers simply let all and everyone dig for themselves, or take any idea as equal to every other. Calvin himself worked out a careful doctrine of accommodation, arguing that the Bible was written for primitive peoples and that it required interpretation. But as Calvinism softened and adapted to the American scene— particularly, as the ideology of the Revolution brought the belief that free people can and must do things for themselves—increasingly American Protestants adopted the position that the Bible is the direct and unambiguous guide for one's life.[2]

Indeed, God himself intended the Bible as just such a guide, and consequently he inspired Scriptures that the most humble could and must read for themselves. Somewhat paradoxically, Jonathan Edwards, whose authority was high at this time, supported this egalitarianism: "It is to be considered that this [the Bible] is given for the rule of all ages; and not only the most learned, and accurate, and penetrating critics, and men of vast inquiry and skill in antiquity." In this he echoed the Apostle Paul: "God has chosen the foolish things of the world to put to shame the wise, and God has chosen the weak things of the world to put to shame the things which are mighty." When it came to interpreting the Bible, no one had the right to dictate. What the Good Book said, the Good Book said. That was the end of the matter. To think otherwise was not just un-Christian, it was un-American.[3]

It also went against Scottish common sense philosophy, which had developed in direct response to the skepticism of the eighteenth century and which underpinned a great deal of the theological thinking in the new republic. In the words of a somewhat younger Henry Ward Beecher than we met in the last chapter: "Where you have had a Bible that the priests interpreted, you have had a king: where you have had a Bible that the common people interpreted; where the family has been the church; where father and mother have been God's ordained priests . . . there you have had an indomitable yeomanry, a state that would not have a tyrant on the throne, a government that would not have a slave or a serf in the field."[4]

More could be said, particularly about the unrestrained culture of printing in the early years of the republic and the many editions of the Bible that became readily available to all. The Scriptures were no longer expensive items, chained to the lectern in church. But the point is made—perhaps made too well. If by the mid-nineteenth century the primacy of the Bible was universally accepted in America, why today are not all Americans Protestants, completely convinced of the literal truth of Genesis? Why are they not all creationists?

One answer is that no one is ever truly a literalist. All reading is an act of interpretation, reflecting the culture of the day. Seventeenth-century Puritans would not have found justification in the Bible for the eighteenth-century Constitution's separation of church and state. And few Christians other than minority groups like the Quakers took seriously the prohibitions against physical retaliation in the Sermon on the Mount. Violence was, indeed, almost as much an American tradition as religion. And a society that considered monetary success in business dealings to be the highest expression of political freedom was hardly likely to take literally Jesus' response to the young rich man who asked the way of salvation: "Go your way, sell whatever you have and give to the poor."[5]

But a more immediate reason explains why biblical literalism came under attack, and Beecher's sermon gives the game away.

Slavery

America had slaves in its fields—black slaves, imported from Africa. By mid-century many could see this as a grave moral failing, a cancer in the nation. Slavery was wrong. Yet—a big yet—the Bible taken literally condones slavery, even endorses it. The patriarchs had slaves, and God did not condemn the practice. Far from telling Abraham to free his slaves, God told him to circumcise the males, whether they wanted this or not. At the other end of the Bible, Paul does not tell slaves to rebel or masters to free; he tells slaves to obey their masters! Jonathan Edwards, a century earlier, had kept slaves and saw no biblical impediment to the practice.

Of course, one can read the Bible as condemning slavery. The commandment to love thy neighbor as thyself argues against forcing a man to labor in the stinking hot fields of Alabama or Mississippi against his will. "Thou shalt not commit adultery" weakens the case for using a slave woman as one's own personal concubine. Nowhere in the Sermon on the Mount can one find justification for splitting up and selling a family to make a buck. But this all demands interpretation and biblical exegesis, as well as the realization that the amateur is not more qualified than the professional in this pursuit.

The slavery issue ripped the nation apart theologically and, with the outbreak of the Civil War, tore it apart politically and socially as well. Northern Protestants, particularly those with roots in older denominations, set out on a new course. In their view, the triumph of the North over the South—clearly God's intention—showed that the strict literalism of the old days was dead, indeed it was positively wrong. "The Bible, then, does not teach untruth. But it cannot without a perpetual miracle eliminate the ignorance, misconception, erroneous implication, and suggestion, the blended truth and untruth, that enter into the very fabric of human speech."[6] People like Beecher were freed to take on more modern, European notions of biblical scholarship. They were also free to enjoy the new prosperity that was released in America after the war, and the concomitant notions of progress. They turned the new sciences, especially evolution, into paving blocks, rather than roadblocks, on the way to salvation.

Northern Catholics fought for the Union the same as Protestants, but generally American Catholicism did not support abolition. The Church was moving toward the philosophy of Thomas Aquinas, who himself owed a debt to the philosophy of Aristotle, with its explicit defense of slavery. Even John Henry Newman—no great friend of Thomism—confessed that, although he disliked the practice of slavery, he did not find the institution "intrinsically evil."[7]

After the war, white southern Protestants felt very differently from their co-religionists in the North. They were a defeated, resentful people, and things were made no better when, after the death of Lincoln, the northerners seemed to be bullying and pushing white southerners around, even gloating. These southerners did not read the Bible as justifying the triumph of the North. Rather, they turned to passages showing that God sometimes lets even his most favored children continue in their suffering. Unflinchingly convinced of the inferiority of blacks, southern whites only intensified their literalist reading of the Bible. It became part of a shell to protect them against the awful North—and to isolate them from the progressive ways of the late nineteenth century.

Although the South may have been the heart of biblical literalism, it did not beat alone. A sophisticated thinker like Beecher might

readily throw off the old ways, but for many in the North and especially in the mid-West and farther out, the Bible continued to have its
established, significant, theological, and cultural role. Many people in
urban enclaves and on the frontier shared the South's discomfort with
modern society—after all, they were having to compete for jobs with
the Irish and other new arrivals from Europe. They were being pushed
into the big new factories, and they were feeling the brunt of capitalism, in good times and even more in bad. In other words, just as
evolution provided reformers of various kinds with either a basis for a
secular religion in its own right or support for a revised Christianity
with a social and cultural role, so the evangelical religion that remained in the North and West likewise played a social and cultural
role. For many, the liberal churches were not speaking to the issues; indeed, the liberal churches were not speaking in Christian terms at all.
Social gospel may have been "social" but it was not "gospel." It was
time to get back to fundamentals.

Against Evolution

And nothing was more fundamental than opposition to evolution.
The greatest American evangelist of the second half of the nineteenth
century, Dwight L. Moody, preached on the "four great temptations
that threaten us today." These were the theater, failure to keep the
Sabbath, Sunday newspapers, and atheism, including evolution. Some
evangelists were blunt in linking enthusiasm for evolution with opposition to slavery—and decrying both sentiments. Pretending that
Adam and Eve were not the sole parents of the race and that time was
not as described in the Bible were "assaults of infidel science upon the
records of our faith, and both have found their warmest advocates
among the opponents of slavery." Not that one had to rely exclusively
on the Bible to oppose evolution. There were always articulate North

American voices ready to speak against it. Louis Agassiz was one; the leading North American theologian, Charles Hodge, of Princeton Theological Seminary, was another. Rhetorically, the title of one of Hodge's books asked: "What Is Darwinism?" Came the stern reply: "It is atheism."[8]

One senses that for Agassiz the evolution issue was not unwelcome. He used his opposition to promote publicly the vision of science that he (and his adoring Boston audience) endorsed and supported. Likewise, the evolution issue was a chance for Hodge to emphasize his traditional religious (but distinctly American) stance. As a Presbyterian, he accepted fully Calvin's position on the Bible. The book is true, completely and utterly true, and binding on the Christian. This had led him into considerable angst and grief over the slavery issue, which the sacred text did indeed seem to support. Hodge only extricated himself at the last (some would have said, beyond the last) moment by some chicanery over whether the slavery of the Bible could ever be identified with the slavery of the American South.[9]

Opposition to evolution was a happier topic, for here Hodge could do justice to the issues. Agreed, God accommodated his language to his audience, and our absolute obligation is to recognize and work with this. "If any man . . . asserts that the Bible teaches that the earth is a plane round which the sun revolves, because for five thousand years men understood the Bible so to teach, he degrades the Word of God & does all he can to undermine its authority." However, accommodation is one thing; craven surrender is quite another. Where revelation leads, reason follows. On natural theological grounds, pure Darwinism is beyond the pale, in Hodge's view. Evidence of design exists and is a definitive argument for final causes. "The doctrine of final causes in nature must stand or fall with the doctrine of a personal God." For Hodge, Darwin's theory of evolution blocks all of this. "Ordinary men reject this Darwinian theory with indignation as

well as with decision, not only because it calls upon them to accept
the possible as demonstrably true, but because it ascribes to blind, un-
intelligent causes the wonders of purpose and design which the world
everywhere exhibits; and because it effectually banishes God from his
works." No Darwin. No evolution.[10]

Calvinists had always wrestled with the need to take science seriously. As a youth, Jonathan Edwards had written a detailed study of
spiders. Our task was to discern and use the ways of God the Divine
Lawmaker. This urge to study science was often less of an imperative
for many other Protestant denominations and movements, particularly those that put major emphasis on the ability of the untutored
heart to reach God through a supposedly simple and direct reading of
the holy text. For the average believer, if the Bible was against evolution, that was enough.

But behind this simple scenario was the bigger picture of American
millennialism. In the eighteenth, and continuing into the nineteenth,
century it was very much part and parcel of the belief in American
exceptionalism. God did not just let the Revolution happen and then
forget about America's existence and about his future plans for his
newly chosen people. Far from it. "The Battle Hymn of the Republic" was a paraphrase of Revelation.

> Mine eyes have seen the glory
> Of the coming of the Lord;
> He is trampling out the vintage
> Where the grapes of wrath are stored;
> He hath loosed the fateful lightning
> Of His terrible swift sword;
> His truth is marching on.

We have already seen how, after the Civil War, liberal Christians
turned increasingly to progressivist ideas, which fit in nicely with a

postmillennialist eschatology, especially one that took everything in a nonliteral fashion. "As applied to the development of life upon the earth, the scientist calls it evolution; as applied to the story of mankind we call it history; as applied to God's supreme purpose we call it the development of the Kingdom of God."[11] Admittedly this eschatology became more and more metaphorical as the years went by, but it gave a vision of the future nevertheless. Going this route was hardly an option for conservative thinkers—they rejected both progress and evolution. How then could millennial thinking have been incorporated into their religion?

The answer lies in the other side of the coin: premillennialism—an interpretation of Revelation that foretells Jesus' second coming *before* the millennium rather than after. A conservative thinker did not absolutely necessarily have to be a premillennialist any more than a postmillennialist had to be committed to the great significance of human progress—theologically speaking, Jonathan Edwards inclined toward postmillennialism—but by the beginning of the nineteenth century in both Britain and the United States this alignment was becoming the norm. A premillennialist did not necessarily have to be gloomy and pessimistic and a postmillennialist did not have to be cheery and optimistic—individual human nature has a way of trumping ideology—but again by the beginning of the nineteenth century in both Britain and the United States the pattern was becoming clear. The postmillennialist was working to achieve better times; the premillennialist thought that change would come only with the Second Coming of Christ, when he and the faithful will defeat the forces of evil at Armageddon. Even though postmillennialism (religious or secular) flourished mightily after the French Revolution, premillennialism proved a hardy plant as well and appealed increasingly to conservatives as the nineteenth century drew to a close.[12]

Premillennialism could and sometimes did backfire badly. If Christ is coming again soon, it is tempting to start working out the date, using the visions and figures given in the apocalyptic books of the Bible. Most famous was the calculation of the upstate New York farmer William Miller, who said that the coming would be in the autumn of 1844. Large numbers of people quit bringing in the harvest and gathered expectantly for the great event—which, for obvious reasons, became known as the Great Disappointment.

But just as scientists often will not let a little counter-evidence stand in the way of a good hypothesis, so premillennialists learned quickly to deal with such nonevents. In the case of Miller, some of his followers argued that there had actually been, at the spiritual level, significant events necessary for the (still future) Second Coming. Taking up with Ellen G. White, a powerful figure much given to visions of her own, this group evolved into the still-thriving Seventh-day Adventists, one of a number of denominations into which nineteenth-century Christianity was much inclined to speciate.[13]

Other premillennialists were greatly influenced by a British, sometime-Anglican clergyman, John Nelson Darby (1800–1882), the first leader of the Plymouth Brethren. Darby was a dispensationalist, believing in certain ages (seven is the usual number), broken by great catastrophes or times of tumult, the first of which was Noah's flood. At some point in the future there will be the Great Tribulation, marked by flight of the truly good to heaven (the so-called rapture of the saints), the return of the Jews to Israel, the coming again of Jesus and destruction of the Antichrist, and commencement of the thousand-year period.[14]

The saving feature of this kind of premillennialism was that people were careful not to specify exactly when the rapture and subsequent events were about to commence—indeed, much was made of various biblical passages which warned that we can never know but must always be prepared. "I want to speak this word to the man who would be free from unclean personal sin: the next time temptation comes, fix your mind on the hope of His coming . . . Can I be caught on the verge of that sin, if I am expecting at that very moment Jesus Christ may come?"[15]

Fundamentalism

In America in the aftermath of the Civil War, religious beliefs like these found a ready market. Dispensationalism was taken up with enthusiasm. Thanks to earnest proselytizing at Bible Conferences and through evangelists and their various colleges and institutes, it seeped also into the thinking of those in more conventional denominations—Baptists, Methodists, and others. Its place was really secured at the beginning of the twentieth century with the arrival of the so-called *Scofield Reference Bible* (1909), an edition that adopted Darby's thinking and sold over two million copies. Having come from the fringes of Protestantism, premillennialism became part and parcel of the general conservative reaction against liberal thinking.

Often the position was linked, especially in Moody's preaching, with a form of "holiness" theology. The Holiness movement started at a series of conferences in Keswick, in the English Lake District, and spread to America. The claim was that we are able to achieve a sinless state right now, thanks to God's grace, so long as we concentrate on personal purity. Against the postmillennialists who stressed effort and good works on earth, premillenialists emphasized the state of one's eternal soul (and the souls of others). It was not incompatible with good works—there was no greater saint in the nineteenth century than the premillennialist Earl of Shaftesbury, who earned his halo by protecting women and children from endless factory work and dangerous and degrading tasks in the mines and elsewhere. Without in any sense undercutting the genuine godliness of Shaftesbury's sentiments toward the suffering, his motive always was to lift the physical burdens of individuals through labor reform and education (he was also much involved in schooling for the masses) so that they could concentrate on the state of their souls. He had no interest in social improvement merely for its own sake.

The premillennial focus was on individual salvation and the necessity of turning our empty broken lives over to Christ. Hence, the very peculiar sentiments expressed in one of the hymns that gospel singer Ira Sankey would use to set the tone for Moody's sermons.

> Oh to be nothing, nothing
> Only to lie at His feet,
> A broken and emptied vessel,
> For the Master's use made meet.
> Emptied that He might fill me
> As forth to His service I go;
> Broken, that so unhindered,
> His life through me might flow.

Moody started his career going to "the poor sinner with the Bible in one hand and a loaf of bread in the other." But the ornery sinners used to concentrate on the bread rather than the Word—"and that was just contrary to the order laid down in the Gospel." So he switched. You cannot achieve the impossible; you cannot save the whole world. "I look upon this world as a wrecked vessel . . . God has given me a lifeboat and said to me 'Moody, save all you can.'"[16]

Princeton Theological Seminary, home for many years of Charles Hodge (and distinct from Princeton University, the home of James McCosh), was a leading institute of biblical inerrancy, the philosophy at the center of Old School Presbyterianism. In 1881 Archibald Alexander Hodge, Charles Hodge's son and successor, and Benjamin Breckinridge Warfield made the definitive statement and defense of "the great Catholic doctrine of Biblical Inspiration, i.e. that the Scriptures not only contain, but ARE THE WORD OF GOD, and hence that all their elements and all their affirmations are absolutely error-less, and binding the faith and obedience of men."[17]

While such powerful intellects were important backing for evangel-icals everywhere, the Calvinism of the Princeton theologians kept them from many of the religious excesses of the new age. Indeed, their support was something of a mixed blessing. Warfield was particularly critical of Holiness thinking ("at once curiously pretentious and curi-ously shallow"), and he was probably inclined to accept some form of evolution.[18] Even among the leaders, there were differences. Moody's enthusiasm for Holiness theology separated him from his associates and followers. Although they tended to dispensationalism, Moody was too positive about the possibility of salvation to accept such a gloomy forecast. In that respect he was a paradigm for other soul-sav-ing premillennialists. Many of them went on missions, spending long, dangerous, and arduous years in lands far away—China especially—trying to spread the gospel.

With moral purity came a drive for physical purity. Drink and to-bacco most obviously had to go, but also tea and coffee for some, and in the case of the Seventh-day Adventists even meat.

> First goes the tobacco, most filthy of all,
> Then drugs, pork and whisky, together must fall,
> Then coffee and spices, and sweet-meats and tea,
> And fine flour and flesh-meats and pickles must flee.
>   Oh, yes, I see it is so,
>   And the clearer it is the farther I go.

Happy consequences of this thrust to vegetarianism were the inven-tions of both peanut butter and Kellogg's corn flakes.[19]

As the twentieth century started to roll along, Bible Christians be-came known as fundamentalists—a term based on a series of twelve pamphlets expounding and defending conservative theology, sponsored by "two intelligent, consecrated Christian laymen" and sent to "every pastor, evangelist, missionary, theological professor, theological student, Sunday School superintendent, Y.M.C.A. and Y.W.C.A. secretary in the English speaking world, so far as the addresses of all these can be obtained."[20] Increasingly, opposition rose against evolution. It went against a literal reading of the Bible, or more precisely, it went against the premillennialist reading of the Bible, with the implications of societal doom and rescue possible only when the Savior comes again. Given its associations of postmillennial progressivism—a philosophy being pushed flat out by secular and religious evolutionists—the fundamentalists had about as clear an embodiment of the forces of evil as they could ever expect.

World War I seemed to confirm this dark picture. Social Darwinism had helped to produce this horror. It had reinforced the "might is right" philosophy pushed by the likes of General Friedrich von Bernhardi and supposedly accepted by all Germans; it was apparently a major motive behind the generals' military drive and the atrocities the soldiers were committing. Vernon Kellogg (no relation to cornflakes), the zoologist who had argued so strongly against war, became engaged in relief work in Belgium and rapidly changed his thinking about the non-necessity of war after long-night conversations with the German High Command. "The creed of the *Allmacht* of a natural selection based on violent and fatal competitive struggle is the gospel of the German intellectuals; all else is illusion and anathema." According to this Teutonic philosophy: "This struggle not only must go on, for that is the natural law, but it should go on, so that this natural law may work out in its cruel, inevitable way the salvation of the human species."[21] This must be stopped, even if we have to match violence with violence.

The premillennialists could and did try to tar liberals with responsibility for the vile thinking of the Germans. They scored some nice hits—for instance, linking the liberals' use of higher criticism with other (to them) revolting aspects of German culture. More generally, tensions occasioned by the war became an excuse for broader opposition to what were seen as the forces of modernism—feminism, socialism, urbanism, immigrationism, and much more, forces all linked with evolution. Forces also linked with progress and postmillennial theology. Commentators remark on the "great reversal" in interest in social issues and increasingly conservative attitudes toward philanthropic enterprises, especially those with any state connections.

Yet for all this, and despite the capture of Jerusalem by the British in 1917, which was taken as a highly propitious indication of the predicted return of the Jews to Palestine, generally the war presented a grave problem for premillennialist theology and philosophy. With the onset of war, Kellogg was but one among many liberal Christians who became super-enthusiasts for bloodshed, accusing the premillennialists—with their inclinations toward noninvolvement in social issues—of lack of patriotism. This charge was not quite fair, for the premillennialists certainly were not slow to enlist and support the cause. But there was some truth in this, and it drove the premillennialists to extra efforts to show their patriotism and general concern for human welfare. This primed them for the decade to come.

Prohibition was a case in point. God is notorious for moving in mysterious ways, and one wonder he performed was to bring together forces on both sides of the millennialist divide to fight the evil of alcohol, one side seeing it as a social issue and the other as a personal issue. Theological differences notwithstanding, at the end of the war, the Christian forces were triumphant in forcing a total prohibition on the manufacture and sale of alcohol. With this out of the way, their energies were available for a new squabble, and the premillennialists particularly were eager for a cause—a cause not only close to their theological interests but one that could reaffirm their true American commitment and spirit. Education was an ideal arena, and evolution was the ideal cause against which to fight.

The Scopes Monkey Trial

The populist politician William Jennings Bryan was much too optimistic to be convinced by the gloomy theology of those who expected an imminent Second Coming. Having been in Woodrow Wilson's cabinet until it (in his opinion) succumbed to the spell of social Darwinism, Bryan had devoted his whole life to improving the lot of his fellow Americans, especially the poor, the uneducated, the exploited. But Bryan knew his people, and he knew especially that they resented the modernist philosophy that was now invading the public schools. Particularly in the South, where premillennial theology rode high, opposition to the ways of outsiders was intense. Ironically, this backlash was brought on precisely by the success of the postmillennialist progressionists.[22]

Enrollment in American high schools had shot from 200,000 in 1890 to over two million in 1920. This sparked major stresses and strains. An often-prescribed textbook was *A Civic Biology* by George Hunter. By today's standards the book was pretty mild. It contained material about proper diet, the importance of exercise, and the virtues of going to the lavatory on a regular basis, with many practical tips for the constipated. Alcohol consumption got a fairly rough ride, and, for better or worse, sexuality was unmentioned. The diagrams of the human body left the ignorant uninformed about the human genitalia. The average teenage reader of the book must have been surprised to discover that a civically biological human apparently had the reproductive apparatus of a Barbie Doll.

But evolution did get an unambiguous—if brief—mention, and

natural selection was touched on, if only to suggest that the mechanism is now outmoded. More than this, the discussion was linked to human progress through the need to practice selective breeding—eugenics. In other words, evolution was made part and parcel of schemes for human improvement, biological and social. Biological misfits were "true parasites," but fortunately "we do have the remedy of separating the sexes in asylums or other places and in various ways preventing intermarriage and the possibilities of perpetuating such a low and degenerate race."[23]

Not that people objected to the practice of eugenics as such. No one was that keen on parasites, true or otherwise. Rather, the whole underlying philosophy of using science—especially evolutionary science—to improve humankind was distasteful. It smacked of playing God. At least, this was the position of the good people of Tennessee, a state that had seen high school attendance jump in the second decade of the twentieth century from 10,000 to 50,000. A law was passed proscribing the teaching of evolution, and as a result a young school teacher, John Thomas Scopes, was tried for using Hunter's text in class. Prosecuted by Bryan and defended by Clarence Darrow, who the previous year had saved murderers Nathan Leopold and Richard Loeb from the electric chair, Scopes and his trial caught the attention of the world, especially thanks to the inflammatory reporting of *Baltimore Sun* journalist H. L. Mencken.

Matters descended to the farcical when, denied the opportunity to introduce his own science witnesses, Darrow put the prosecutor Bryan on the stand.

Darrow picked up the Bible and began to read: "'And the Lord God said unto the serpent, Because thou hast done this, thou art cursed above all cattle, and above every beast of the field; upon thy belly shalt thou go and dust shalt thou eat all the days of thy life.' Do you think that is why the serpent is compelled to crawl upon its belly?"

"I believe that."

"Have you any idea how the snake went before that time?"

"No sir."

"Do you know whether he walked on his tail or not?"

"No, sir, I have no way to know."

There was a howl of laughter from the crowd.

Suddenly Bryan's voice rose, screaming, hysterical: "The only purpose Mr. Darrow has is to slur at the Bible . . . I want the world to know that this man, who does not believe in a God, is trying to use a court in Tennessee—"

"I object to your statement." Darrow was contemptuous. "I am examining you on your fool ideas that no intelligent Christian on earth believes."

Judge Raulston put an end to the argument by adjourning the court.

That night, at last, it rained.[24]

A wonderful story. In the play, the Darrow-figure vanquishes the Bryan-figure, yet shows tolerance and understanding when, left alone at the end in the courtroom, he picks up the Bible and the *Origin* and thrusts both into his carrying bag.

As always, things were slightly more complex than fiction would have it. *Inherit the Wind* took many liberties in telling the tale, shaping the clash to reflect concerns of the 1950s rather than the 1920s—specifically the right to have contrary opinions when, thanks to the Cold War, Americans were under great pressure to conform to general norms of behavior and thought. In the real Scopes Trial, evolution—particularly human evolution—was certainly at the center of the debate, but the true issue was more the general philosophy for which evolution was taken to stand. Evolutionism rather than mere evolution. The real Bryan was cautious about dates and ages. In the fictionalized movie about the trial, he was made to look like a buffoon when asked the actual length of the days of creation. In true life Bryan always believed that the days were periods of time.

Bryan: I think it would be just as easy for the kind of God we believe in to make the earth in six days as in six years or in 6,000,000 years or in 600,000,000 years. I do not think it important whether we believe one or the other.

Darrow: Do you think those were literal days?

Bryan: My impression is that they were periods, but I would not attempt to argue as against anybody who wanted to believe in literal days.[25]

After the trial, the general opinion among fundamentalists was that Bryan had acquitted himself well and done a good job. Moreover, while it seems true that the fundamentalist movement peaked with the Scopes Trial and then declined in numbers and influence, it did leave significant lasting effects, especially on education—a topic for a later chapter.

9

Population Genetics

Still leaning against the incubators, he gave them, while the pencils scurried illegibly across the pages, a brief description of the modern fertilizing process; spoke first, of course, of its surgical introduction—"the operation undergone voluntarily for the good of Society, not to mention the fact that it carries a bonus amounting to six months' salary"; continued with some account of the technique for preserving the excised ovary alive and actively developing; passed on to a consideration of optimum temperature, salinity, viscosity; referred to the liquor in which the detached and ripened eggs were kept; . . . [how] this receptacle was immersed in a warm bouillon containing free-swimming spermatozoa . . . how the fertilized ova went back to the incubators; where the Alphas and Betas remained until definitely bottled; while the Gammas, Deltas and Epsilons were brought out again, after only thirty-six hours, to undergo Bokanovsky's Process . . . a bokanovskified egg will bud, will proliferate, will divide. From eight to ninety-six buds, and every bud will grow into a perfectly formed embryo, and every embryo into a full-sized adult. Making ninety-six human beings grow where only one grew before. Progress.

*B*ut of course it isn't. Progress, that is. Rather, we have an absolute nightmare of misguided technology, deadening of the soul as well as (for everyone below Beta) the intellect too. The Director of Hatcheries and Conditioning is a scientist gone mad, able to function only in such a milieu as envisioned in *Brave New World*, the novel by Aldous Huxley, younger brother of Julian and another grandson of Thomas

Henry Huxley. Published in 1932 and from then on a staple of high school English classes, the book described an absolutely horrific future where everything worthwhile and human had been taken and cleansed and mechanized and sanitized—easy, mechanical, passionless sex, readily available drugs to while away the tedious hours, and a quick, painless, fearless death at the end. Nothing stressful. Nothing of value. One finds it hard to imagine a more devastating critique of progressivist ideology. All of that striving, to end up with absolutely nothing worth cherishing.

After two hundred years, progress was truly under attack. Aldous Huxley was not alone in his pessimism about the ability of progress to deliver on its promises, and there were grounds for such a sad picture of the future. World War I had torn Europe apart, killing and maiming millions of the brightest and bravest young men. Then, after the false dawn of the 1920s came the fall of the stock market and the dismal summer of the 1930s, when the Great Depression threw half the world out of work and brought poverty and starvation and despair to more than could be numbered. Totalitarian regimes—Russia under Stalin, Germany under Hitler, Italy under Mussolini, later Spain under Franco—arose in part in reaction to all of this. Surely now was the time when progress of all kinds—cultural and biological—would vanish from the world. Evolution could no longer be, as before, merely a vehicle for social and quasi-religious beliefs. A nonprogressive evolution was about to emerge. But the story is far from simple.

Among the sciences, professional evolutionary studies were second-rate and going nowhere. All of the bright young men at the end of the nineteenth century came to realize this. William Bateson, as an old man remembering the days when he started out, wrote that he and his fellows all became morphologists in order to unlock the secrets of evolution; and for morphologists, embryology was at the cutting-edge. "Therefore every aspiring zoologist was an embryologist, and the one topic of professional conversation was evolution." But it did not work. "Discussion of evolution came to an end because it was obvious that no progress was being made. Morphology having been explored in its minutest corners, we turned elsewhere." People like Bateson wanted to work in the life sciences as trained professionals, experimenting, observing, predicting, explaining. Germanized phylogeny tracing was not enough.[1]

So they moved to other areas, such as the study of the cell (cytology) or, in Bateson's case, the study of heredity, which came to be known as genetics. The big breakthrough came from the work of a Moravian monk, Gregor Mendel, whose experiments with peas in his monastery garden in the 1860s revealed the fundamental laws governing heredity. In *Experiments with Plants* he argued that the factors controlling the production of physical characteristics were passed on, unchanged, from one generation to the next. This concept would become known as "particulate inheritance"—inheritance through unchanging "particles," later called genes. Mendel died in obscurity in 1884, but when his work was uncovered at the beginning of the twentieth century, a physical explanation for Mendel's results was soon forthcoming, and Mendelian genetics was fused readily with the science of the cell.

In the second decade of the last century, Thomas Hunt Morgan and his young associates at Columbia University in New York put everything together and articulated what came to be known as "the classical theory of the gene." Biologists were then well launched on a research program from which new ideas and discoveries continued to pour forth, especially after 1953, when James Watson and Francis Crick deciphered the double-helical structure of the deoxyribonucleic acid (DNA) molecule, with which the classical gene is identified.

The theory of heredity as such deals with the stability of genes as they are passed from an individual organism to its offspring. But stability does not lead to evolution. To understand how whole new species come into being over time, geneticists had to uncover the ways in which genes become altered, or "mutate." Investigators found that although genes are stable for the most part, sometimes they do move spontaneously from one form to another. This change was not uncaused, and the unraveling of precise reasons for mutation was a scientific triumph. Asa Gray was proven wrong: mutation never occurs to order just because an organism suddenly needs to adapt to a change in its environment. Mutation is random in this sense. However, it does occur at regularly quantifiable rates; so in another sense it is not really random at all.

Many early Mendelians were drawn to various versions of saltationism—the theory that huge genetic mutations (jumps) turn organisms from one species into another species. Before long, however, it became clear that the large-scale mutations the early Mendelians documented in the laboratory were atypical. In life, many mutations have very small, almost imperceptible effects, and these variations are the raw material on which selection operates. By combining natural selection with genetics, biologists hoped to update the theory of evolution that Darwin presented in the *Origin*.

The first order of business was to understand the flow of genes throughout a population.[2] Natural selection can bring about change only if some members of a population survive and reproduce and in the process pass along their genetic variations, while the rest do not. The key theorem here was the Hardy-Weinberg law, which demonstrated that unless outside forces impinge on a group, the genes within it will stay in equilibrium, no matter what the initial ratios—analogous in some sense to Newton's first law of motion. What was needed were people who could quantify these outside forces—random mutation, immigration and emigration, selection pressures, and so forth—

and show how the biological equivalent of Newtonian mechanics could be articulated and developed. The two most important and influential workers in this field were Ronald A. Fisher, a statistician in England, and Sewall Wright, an animal geneticist in America.

Ronald A. Fisher

Fisher was a genius. Everyone accepted that. He was a brilliant mathematician, made even more remarkable by weak eyesight, which forced him to do almost all of his thinking without visual aids. He trained at Cambridge as a mathematician and theoretical physicist and worked at a plant-breeding station helping to interpret results and design experiments. Modern statistics virtually dates from the work he did at this point in his career.

But calculation for its own sake was never Fisher's driving passion. He was a deeply conservative Englishman, believing that the ways of his native land were far superior to all others, and as part of this bias he worshiped the memory of Charles Darwin (bolstered by a very close relationship with Charles Darwin's youngest son, Major Leonard Darwin). Fisher wanted to pick up and bear the evolutionary burden for his generation of scientists, and in a paradoxical way—having not been exposed to the phylogeny tracing of the fulltime evolutionists or the antiselectionism of the Mendelians—his nonbiological background and sophisticated training in mathematics prepared him to take a much more purely Darwinian perspective on evolutionary change. In 1930 Fisher published his masterpiece, *The Genetical Theory of Natural Selection*, a work that many to this day regard as second only, in evolution's history, to the *Origin of Species*.[3]

Fisher began with a detailed discussion of the way in which selection can be melded with a theory of particulate inheritance—indeed, how selection really must be melded with such a theory—and how then the central propositions of what came to be known as population genetics can be derived. With this basic background in place, Fisher was ready to present his own picture of the evolutionary process. Overall, he saw populations as trying to push upward to a kind of adaptive excellence, but always having trouble doing so because of changes in the environment. This idea of upward change was captured in his so-called fundamental theorem of natural selection: "The rate of increase in fitness of any organism at any time is equal to its genetic variance in fitness at that time."[4]

Fisher likened this law to the second law of thermodynamics (entropy always increases), but whereas the second law describes degeneration, Fisher's fundamental theorem described biological progress. Fisher saw selection as a gradual, subtle process that nudges and guides rather than forces. "I think of the species not as dragged along laboriously by selection like a barge in treacle, but as responding extremely sensitively whenever a perceptible selective difference is established. All simple characters, like body size, must be always very near the optimum."[5] The Hardy-Weinberg law described the conditions for equilibrium, but real life consisted of constant flow, in the long run making for progress, and not a state of equilibrium.

With the fundamental theorem introduced and explained, Fisher tackled a range of evolutionary topics. Typical was a problem that had puzzled Darwin himself: why sex ratios remain constant and equal under normal circumstances. Darwin's first insight in the *Descent* had been correct, but then he equivocated. Fisher was able to show in detail that balance follows from the act of selection. If males are less common in a group, then selection pressure will produce more males than females, up to the point where they are in balance. The converse happens when females are rare. The emphasis here is on selection for the individual and not for the group. Given that males usually do little or nothing in raising offspring, a group might be better adapted if it had (let us say) 10 percent males instead of 50 percent. But selection does not work that way. Except in special circumstances (as when the competing males are all very closely related), selection will work to raise the number of males. Why? Because if there are fewer males, say just one in ten, then proportionately a parent with a son will have more grandchildren than a parent with a daughter. And any mutations that might have favored the production of a son (at a time when most people were producing daughters) will be passed along disproportionately to the next generation, until the sex ratio is balanced.

Fisher offered much more along these lines, including a detailed defense of Batesian mimicry. But the theme throughout the book was that selection works on random variation, leading to adaptive change.

Sewall Wright

The other major figure, Sewall Wright, was from the American mid-West, studied at Harvard, and then joined the United States Department of Agriculture, focusing on cattle breeding.[6] In 1925 Wright moved to the University of Chicago, and around this time he started to write on evolutionary questions. After Fisher published his book, Wright followed suit with his ideas. And very different ideas they were from those of Fisher. The formal mathematical conclusions were the same, and the Hardy-Weinberg law was the background for both men. Their shared aim was to see how different causal factors can affect and change gene ratios. But beyond this, the divergence was nigh absolute.

Fisher always saw natural selection as the determining factor in evolutionary change. For him, change was a matter of random mutations coming into large populations and either failing to make the grade or producing variations superior to those already existing and so rapidly spreading throughout the group. For Wright, what really counted was homogeneity and heterogeneity and the interplay between the two: "Evolution as a process of cumulative change depends on a proper balance of the conditions, which, at each level of organization—gene, chromosome, cell individual, local race—make for genetic homogeneity or genetic heterogeneity of the species. The type and rate of evolution in such a system depend on the balance among the evolutionary pressures considered here."[7]

Influenced by his time in agriculture, where he had studied successes with shorthorn cattle—the breeders isolated good strains and in-bred them severely—Wright argued that the key to evolutionary change is the fragmentation of large populations into small groups (presumably as a result of external causes). It is in these small subpopulations that real innovation takes place, according to Wright; selection pressure (beyond whatever external force separated the small group in the first place) is irrelevant. Genetic change is random, a result simply of the vagaries of breeding. Although one form A may be biologically superior (fitter) to another form B, random mating in a small group, which will invariably have some coincidences and distortions, might mean that B will prevail nevertheless. Its genes will "drift" to being the only ones represented in the subpopulation. Later, as barriers break down and the subpopulation rejoins the main body of the group, these new B features, which themselves may not have been particularly adaptive when they first arose, may turn out to have advantages by the time the population reunites. If so, they will become fixed within the whole group through selection pressure. Wright's point was that adaptive features have to come into existence in the first place, and isolation and drift—not selection—brings this about.

Wright called his theory the shifting balance theory of evolution. In support of his ideas, he introduced the powerful metaphor of an adaptive landscape, with successful species at the tops of peaks and failures down in the valleys. Evolution moves species from one peak to another.

> Let us consider the case of a large species which subdivided into many small local races, each breeding largely within itself but occasionally crossbreeding. The field of gene combinations occupied by each of these local races shifts continually in a nonadaptive fashion. With many local races, each spreading over a considerable field and moving relatively rapidly in the more general field about the controlling peak, the chances are good that one at least will come under the influence of another peak. If a higher peak, this race will expand in numbers and by crossbreeding with the others will pull the whole species toward the new position. The average adaptiveness of the species thus advances under intergroup selection, an enormously more effective process than intragroup selection. The conclusion is that subdivision of a species into local races provides the most effective mechanism for trial and error in the field of gene combinations.
>
> It need scarcely be pointed out that with such a mechanism complete isolation of a portion of a species should result relatively rapidly in specific differentiation, and one that is not necessarily adaptive.[8]

Wright did not want to deny that natural selection operates and that organisms develop adaptive features that are significant to their survival. But he thought that many features develop that simply have no effects on survival and reproduction. Apparently only when you start to get up to the subfamily or family level do you start to get adaptive difference. "The principal evolutionary mechanism in the origin of species must thus be an essentially nonadaptive one."[9] In other words, one has lots of nonadaptation, but even for deeply adaptive characteristics—especially very innovative ones—selection does not play a vital role in their creation.

Professionalism and Progress

The work of Fisher and Wright and their fellows really was a watershed in the history of evolutionary theorizing. They were professional scientists with professional standards. Work had to be consistent and coherent with other sciences, it had to be mathematicized if possible, it had to lay itself open to check and offer possibilities of prediction, and more. These were the criteria by which they expected their evolutionary theorizing to be judged, and by which they judged the theories of their contemporaries. Darwin's dream of a professional science of evolution was at last on its way to being realized. Does this then mean that cultural values, specifically progress with its odor of postmillennialism, had vanished? And if they were gone, was it because such values were incompatible with professional science, or because progress was now no longer a cherished value? Did this mean that biological progress was no longer a factor in evolutionary thinking?

Nothing could be further from the truth. Both Fisher and Wright were ardent cultural progressivists and enthusiastic biological progressivists. Fisher's evolutionary progressivism was combined with a deep commitment to Christianity. He saw God as having a task here on this earth, namely, creating human beings, creatures made in his own image, which he does through the fundamental force of natural selection. In the living world as opposed to the physical world, an entropy-reversing process leads ultimately to the highest of all possible products, human beings. These human beings, likewise, have a task of improvement here on earth. God's creation was not an instantaneous, one-off event but a long process that takes place over time and is still in progress. "In the language of Genesis we are living in the sixth day, probably rather early in the morning, and the Divine artist has not yet stood back from his work, and declared it to be 'very good.' Perhaps that can only be when God's very imperfect image has become more competent to manage the affairs of the planet of which he is in control."[10]

This thought supported a remarkable parallel between faith/works and Lamarckism/Darwinism:

> There is indeed a strand of moral philosophy, which appeals to me as pure gain, which arises in comparing Natural Selection with the Lamarckian group of evolutionary theories. In both of these contrasting hypotheses living things themselves are the chief instruments of the Creative activity. On the Lamarckian view, however, they work their effect by willing and striving only; but, on the Darwinian view, it is by doing or dying. It is not mere will, but its actual sequel in the real world, its success or failure, that is alone effective.
>
> We come here to a close parallelism with Christian discussions on the merits of Faith and Works. Faith, in the form of right intentions and resolution, is assuredly necessary, but there has, I believe, never been lacking through the centuries the parallel, or complementary, conviction that the service of God requires of us also effective action. If men are to see our good works, it is of course necessary that they should be good, but also and emphatically that they should work, in making the world a better place.[11]

What is our task here on earth? Fisher, a life-long member of the Church of England, focused on eugenics. Specifically, he feared that as culture developed, the upper classes (that is, the genetically superior) would reproduce less and less. Given the over-reproduction of those in the lower ranks, the result would be biological degeneration and an end to God's progress. Our task is to reverse this, and as a mark of personal commitment he himself married a young girl, chosen especially for her breeding potential, and then proceeded to have eight children, despite the fact that at times his family was reduced to near poverty.

Sewall Wright was no less of a progressionist, and his vision of the upward process was as much a function of his culture as was that of Fisher. Wright was no Christian, although he was drawn to (and for many years worshiped with) the Unitarians. He was, however, intensely, obsessively, interested in philosophical issues of a metaphysical kind. He was always worrying away at problems about the nature of reality and our relationship to ultimate being. In particular, Wright harbored the very odd belief that everything—everything!—had consciousness, from molecules to men, and that perhaps we are on the way up to a kind of super-organism, with super-consciousness. This "panpsychic monism" was deeply progressionist, hierarchically and temporally.

> The greatest difficulty is in appreciating the possibility of the integration of many largely isolated minds into a higher unitary field of consciousness such as must necessarily occur under this viewpoint in the organism in relation to its cells; in these in relation to their molecules and in these in relation to their molecules and these in relation to more ultimate entities. The observable hierarchy of physical organization must be the external aspect of a hierarchy of mind.[12]

Odd beliefs, but in the 1930s far from unique. The much-read and greatly admired novel, *Last and First Men: A Story of the Near and Far Future*,

by Olaf Stapledon, posited future races of ever-more intelligent beings. The final breed had precisely such a kind of shared consciousness, one that was supposed to make serious intragroup conflict impossible.

> Occasionally there is a special kind of group intercourse in
> which, during the actual occurrence of group mentality, all the
> members of one group will have intercourse with those of an-
> other. Casual intercourse outside the group is not common, but
> not discouraged. When it occurs it comes as a symbolic act
> crowning a spiritual intimacy.
>
> Unlike the physical sex-relationship, the mental unity of the
> group involves all the members of the group every time it oc-
> curs, and so long as it persists. During times of group experi-
> ence the individual continues to perform his ordinary routine of
> work and recreation, save when some particular activity is de-
> manded of him by the group-mind itself. But all that he does as
> a private individual is carried out in a profound absent-minded-
> ness. In familiar situations he reacts correctly, even to the extent
> of executing familiar types of intellectual work or entertaining
> acquaintances with intelligent conversation. Yet all the while he
> is in fact "far away", rapt in the process of the group-mind.
> Nothing short of an urgent and unfamiliar crisis can recall him;
> and in recalling him it usually puts an end to the group's experi-
> ence.[13]

From all accounts, Sewall Wright was a rather boring man, whose
idea of a fun night out was a metaphysical discussion with his buddies
down at the local faculty club. The fantasy of mental group inter-
course was probably beyond his imagination, but the idea of group
consciousness was there. What were the ideas that led up to it? Just as

the fundamental theorem gave the answer to Fisher, so the shifting balance theory gave the answer to Wright. Groups become isolated, some of these subpopulations scale new adaptive peaks, and then the whole population gets dragged upward. In part, the influence driving Wright's thinking was the shorthorn experience. But another factor lurked beneath the surface, and that was the influence of Spencer. The shifting balance theory, with its stability and disruption, its random creation of new innovations by drift, followed by a return to stability, was not so much Darwinian as Spencerian. It was a version of his dynamic equilibrium.

This might be expected of a young American intellectual of the early part of the twentieth century. As a graduate student at Harvard, Wright was exposed to Spencerian influences, most significantly from his teacher, the chemist Lawrence J. Henderson, author of the well-known work *The Fitness of the Environment* and a Spencer fanatic. Henderson's writings exhibited all sorts of organismic analogies, upward movement, changes from simplicity to complexity, and, most prominently, that ever-changing flow to and from a state of balance. As Henderson said explicitly: "Spencer's belief in the tendency toward dynamic equilibrium in all things is of course fully justified." The student was brought under the spell: "I was always very much impressed with Henderson's ideas."

To his brother Quincy, Wright acknowledged explicitly the influence through Henderson back to Spencer: "I found him a very stimulating lecturer and got lots of ideas from him, 'condition of dynamic equilibrium' etc." In another letter, he toyed with a classificatory scheme that was going to include the whole of nature ("electron, atom, animal, etc.") as a tension between equilibrium and "Change of equilibrium (Evolution)," where the latter covers the "Mechanism of change." In this same letter, Wright described genetics as dealing "with both the maintenance of equilibrium in the species (heredity) but also the mechanisms of change in this equilibrium (variation by recombination of factors and otherwise)."[14]

How were Fisher and Wright able to get away with it? They were producing professional-quality, evolutionary biology and yet they were as much influenced by progress as Lamarck, Chambers, or Spencer. The answer is that they tried to have things both ways. In principle, the fundamental theory and the adaptive landscape do not have to be interpreted in a progressionist fashion. The upward path can get reversed. As landscapes change, everything might get taken back down to a lower point. We humans could become Eloi or Morlocks. So neither Fisher nor Wright made cultural values a necessary part of their evolutionary picture.

But this said, if one wanted to continue with a progressionist evolution, the opportunities were there to be grasped. For the English, it meant a kind of souped-up Darwinism, with progress really incorporated into the causal picture in a way that eluded Darwin himself. For the Americans, it meant a kind of souped-up Spencerianism, with progress really incorporated into the causal picture in a way that Spencer beyond the grave must have applauded. From Fisher's and Wright's theories of population genetics, it was but a few easy steps to a metaphysics of evolutionism.

The Empiricists

After the theoretical population geneticists had done their creative work, the empiricists—the naturalists and experimenters—moved in to put flesh on the formal skeletons. And here the story takes an interesting divide. Both Fisher and Wright had their chief lieutenants, men who worked with them and translated their ideas into functioning theories and research programs. But the way in which this translation occurred was very different in England and America. Fisher's empirical disciple was Edmund Brisco ("Henry") Ford, an Oxford-trained biologist who worked on butterflies and founded a school that he labeled ecological genetics.

Ford worked with Fisher on several seminal papers and was brilliant at discipline building. Under his guidance, the 1950s saw outstanding work by a group of young men, most notably Arthur J. Cain and Philip M. Sheppard, on such problems as the shell-banding of snails and the effects of predation by birds. Phenomena that had hitherto been ascribed to random factors were shown to be tightly under the control of selection. At the same time, Ford was talented at fundraising. He persuaded the Nuffield Foundation, a private philanthropic body, that the work he and his associates were doing on fast-breeding invertebrates was an excellent model for human beings. Ecological genetics could throw valuable and needed light on human disease and its transmission.

But one thing that Ford shunned absolutely was progress. He saw that any hint of progress in the new science he was founding would be fatal. Evolution was his field, not evolutionism. He had seen firsthand the damage that progress could do to serious science. The most public scientist in Britain at that time was Julian Huxley, and he was forever linking progress with evolution—in prose and in verse. Thus the (truly dreadful) sonnet, "Man the Philosophizer."

> Amoeba has her picture in the book,
> Proud Protozoön!—Yet beware of pride.
> All she can do is fatten and divide;
> She cannot even read, or sew, or cook . . .
> The Worm can crawl—but has no eyes to look:
> The Jellyfish can swim—but lacks a bride:
> The Fly's a very Ass personified:
> And speech is absent even from the Rook.

The Ant herself cannot philosophize—
While Man does that, and sees, and keeps a wife,
And flies, and talks, and is extremely wise . . .
Yet our Philosophy to later Life
Will seem but crudeness of the planet's youth,
Our Wisdom but a parasite of Truth.

Julian Huxley's mother was the niece of Matthew Arnold—proof that not all change is progress. This did not stop Huxley from harping on about the topic without cease, even in his major scientific work, *Evolution: The Modern Synthesis*. Yet here things backfired. As far as professional scientists were concerned, if Huxley wanted to follow in the tradition of Erasmus Darwin, hymning evolutionary progress in verse, that was his business; writing poetry is a fairly harmless activity. If Huxley wanted to promote evolutionism on radio and television— he was a weekly performer on a very popular general-knowledge show, the *Brains Trust*—then so be it. But when Huxley tried to carry this over to the realm of professional science, he met derision and opposition. The serious scientists, especially molecular biologists, would have nothing of it. They sneered at Huxley and cut him out of the loop for grants and other support.[15]

Ford saw this writing on the wall and took note. For all he was a devoted Fisherian, he cut the fundamental principle out of his own presentation of evolutionary thought. Starting with a little text that Ford wrote back as early as 1931, the fundamental theorem simply went unmentioned, as Ford moved on to topics like sex ratios and mimicry.[16] Fisher had tried to remold Darwinism so that it had an inherently progressivist core, and Ford would have none of this. English professional evolutionary biology was to be culture-value-free.

America went a somewhat different route. Wright's disciple was the Russian-born geneticist Theodosius Dobzhansky. He took over the shifting balance theory and made it the center of his evolutionary thinking, especially as expounded in his major treatise *Genetics and the Origin of Species.* Following Wright, Dobzhansky saw really important evolution involving the fragmentation of populations, change within subgroups, and then new characteristics reaching out and moving through the whole population. Genetic drift was a key player in Wright's world, and the same was true for Dobzhansky. He was not convinced that selection played the same role in adaptation that people like Fisher assumed. To get new groups, all you needed was the chance breakup of a group and enough time for mutations to accumulate. "This statement is not to be construed to imply a denial of the importance of selection. It means only that racial differentiation need not necessarily or in every case be due to the effects of selection."[17]

The fact is that *Genetics and the Origin of Species* was not a very Darwinian book. Dobzhansky simply was not obsessed with the problems that preoccupied Darwin, Fisher, and their followers. The best evidence for natural selection to date—mimicry—was discussed, but the treatment was cool and noncommittal. Indeed, for anyone schooled in the tradition of Bates and Fisher, the indifference was positively jaw-dropping. "Taken as a whole, an unprejudiced observer must, I think, conclude that an experimental foundation for the theory of protective resemblance is practically non-existent."[18] Yet, shortly after *Genetics and the Origin of Species* first appeared, Dobzhansky's somewhat distant attitude toward selection and adaptation changed quite significantly.

Using the fruit fly as his model organism in nature as well as in the laboratory, Dobzhansky set up a long series of studies of variation and change in evolution.[19] He soon discovered that what he and his fellows had taken to be paradigmatic examples of drift—variations in chromosome structures within and between populations—simply had to be interpreted in terms of natural selection. Some seasonal variations shared by isolated populations of fruit flies in the wild, which move from one chromosome form to another and then back again, cannot occur purely by chance and must be ascribed to adaptive advantage—one chromosome form being favored at one season and another form at another season. Soon all of this was confirmed by simulated changes occurring in experimental populations kept in the laboratory. Natural selection was a lot more pervasive and powerful than Dobzhansky had realized.

His new-found respect was reflected in the third edition of *Genetics and the Origin of Species*, published in 1951. Drift existed but its role was much reduced, while evolution was now a lot more adaptive and selection-driven. At another level, things were quite unchanged. The adaptive landscape metaphor was as important as ever, but the stress now was on the extent to which selection plays a role in putting organisms on the peaks and keeping them there. Homogeneity and heterogeneity and upward rise were far from gone.

The Synthetic Theory of Evolution

Dobzhansky was no less gifted than Ford in discipline-building. He gathered students and received funds from government agencies of one sort or another. For example, he persuaded the Atomic Energy Commission that evolutionary studies throw useful light on the effects of nuclear-weapons testing. Like Ford, Dobzhansky and his followers realized that any hint of cultural values as a driving force behind their work would be fatal to its prospects for becoming a respected professional discipline. They set out deliberately to purify their own work and anything to which they gave their seal of approval. As part of their campaign to revitalize evolutionary studies, they founded a new journal, *Evolution*.

Ernst Mayr, a German-born ornithologist, was the first editor, and his letters to prospective contributors made very clear the need to stay away from dangerous ideas. As he wrote to one would-be author, "The field has reached a point where quantitative work is badly needed. Also, evolutionary research, as you realize, has shifted almost completely from the phylogenetic interest (proving evolution) to an ecological interest evaluating the factors of evolution."[20] To others, Mayr was explicit that any talk about progress, or explicit reliance on theories that promoted progress (like orthogenesis), were strictly *verboten*. "Your manuscript, 'Orthogenesis in Evolution,' has been studied by two readers of the Editorial Board who have reported back to the Editor that they do not consider the manuscript suitable for EVOLUTION."[21] In another letter he pointed out that "the prestige of evolutionary research has suffered in the past because of too much philosophy and speculation."[22]

Was the progressivist coloring of Wright's theory a deep embarrassment to these American supporters of the synthetic theory of evolution (as they labeled their project)? Had progress become a phylogenetic relic in science, like the appendix? Absolutely not. To a person, all of the new, professional, American evolutionists were ardent progressionists, and for most of them that was precisely why they had been attracted to evolutionary studies in the first place. Like Cuvier over a hundred years before, they realized that for professional reasons they had to play the game of being culture-value-free, otherwise there would be no grants, no prestigious university posts, no students, no respect. Evolution was their profession. But evolutionism was their obsession.

Their strategy was clever and simple. They would publish two sets of books. One professional, with no hint of progress. One popular, with much talk of progress. Two messages, for two audiences. Most instructive was the example of the paleontologist George Gaylord Simpson. In 1944 he published *Tempo and Mode in Evolution*, in which he applied population genetics, specifically Wright's shifting balance theory, to the fossil record. The work was spartan in its avoidance of anything extrascientific. In 1953 he published a revision of this work, with the new title *Major Features of Evolution*. Again, not a hint of progress or of any other cultural value. But in between, in 1949, Simpson delivered a set of public lectures at Yale University which were collected into a book, *The Meaning of Evolution: A Study of the History of Life and of Its Significance for Man*. Biological progress was the connecting thread throughout, and the basis of life's major social directives.

The first premise of Simpson's evolutionary ethics was that knowledge was a good in itself that should be promoted. "As a first proposition of evolutionary ethics derived from specifically human evolution, it is submitted that promotion of knowledge is essentially both the acquisition of new truths or of closer approximations to truth (metaphorically the mutations of the new evolution) and also its spread by communication to others and by their acceptance and learning of it (metaphorically its heredity)." The second ethical principle that Simpson derived from evolution was personal accountability, which leads to integrity and dignity. "The responsibility is basically personal and becomes social only as it is extended in society among the individuals composing the social unit. It is correlated with another human evolutionary characteristic, that of high individualization."[23]

This valuing of responsibility and dignity and so forth was very much a function of the times and society within which Simpson lived. The cold war had settled into a long winter, and Soviet science was under the thumb of Lysenko, a man who openly used his power to oppress (even unto death) those biological opponents who dared to point out how fraudulent his work truly was. For biologists like Simpson, this persecution of genuine scientists made the pursuit of freedom and democracy a personal crusade. From evolutionary speculations about dignity and responsibility, Simpson launched straight into a condemnation of the oppressive regimes then flourishing in the East, and he juxtaposed this with a cherishing of the society within which he found himself: "Democracy is wrong in many of its current aspects and under some current definitions, but democracy is the only political ideology which can be made to embrace an ethically good society by the standards of ethics here maintained."[24]

Here we have the testament of a secular postmillennialist. By 1959, the centennial of the publication of the *Origin of Species*, evolutionary biology had become professional. Yet, one would be naive to conclude that evolutionary thinking had ceased to play its traditional role as a secular religion. In the New World particularly, Herbert Spencer was a more pervasive and ongoing influence than anything that had come from the village of Downe.

10

Evolution Today

I didn't know, nor have I ever discovered, who let go first. I'm not prepared to accept that it was me. But everyone claims not to have been first. What is certain is that if we had not broken ranks, our collective weight would have brought the balloon down to earth a quarter of the way down the slope a few seconds later as the gust subsided. But as I've said, there was no team, there was no plan, no agreement to be broken. No failure. So can we accept that it was right, every man for himself? Were we all happy afterwards that this was a reasonable course? We never had that comfort, for there was a deeper covenant, ancient and automatic, written in our nature. Co-operation—the basis of our earliest hunting successes, the force behind our evolving capacity for language, the glue of our social cohesion. Our misery in the aftermath was proof that we knew we had failed ourselves. But letting go was in our nature too. Selfishness is also written on our hearts. This is our mammalian conflict—what to give to the others, and what to keep for yourself.

*T*his internal monologue comes as part of the dramatic opening to the novel *Enduring Love* by the British author Ian McEwan. A science writer, Joe Rose, is having a picnic somewhere in the English countryside with his live-in partner, Clarissa. They see a hot-air balloon in trouble, and Joe rushes over to try to help. A number of other people are doing likewise. These would-be helpers include a doctor in his early forties, John Logan. Unfortunately, the balloon starts to rise up with a small child in the bottom of the basket. All let go except

Logan, who is carried high into the air at the end of a rope. Finally, he falls several hundred feet and is killed.

McEwan is hardly the first novelist to wrestle with the conflicts between self-interest and altruism—between our own desires and that totally selfless giving that is captured in the novel's title, that feeling for others described by Saint Paul in his first epistle to the Corinthians: "Love bears all things, believes all things, hopes all things, endures all things." But McEwan is the first novelist to express the dilemma in the terms of modern evolutionary biology. Cooperation versus selfishness—"our mammalian conflict." This formulation seems a long way from the Tennysonian "nature red in tooth and claw," or the Darwinian struggle for existence, or the Spencerian laissez faire. Has modern evolutionary theory broken from its past?

The last fifty years have witnessed the ongoing success of a professional evolutionary biology—a discipline that has used Charles Darwin's natural selection profitably to explain organic change.[1] Adaptation is now regarded as the chief characteristic of living matter, and this assumption or hypothesis has been the guiding tool of evolutionary researchers. A field of study has taken root in universities, either within established biology departments or on its own (often allied with ecology). It has students, academic organizations, publications, and conferences—all the hallmarks of a successful branch of science. It suffers still from competition with more obviously fundable areas of biological science, notably those that investigate the molecular aspects of organisms. But even here there have been advances and fertile collaborations, as molecular biologists have recognized that evolution can raise interesting problems and suggest fruitful solutions, and as evolutionary biologists have realized that both the methods and concepts of molecular biology can be invaluable tools of evolutionary research.

Although the evolutionary biologists of fifty years ago were wary of the brash, all-conquering triumphs of those men (and a few women) working on the molecular biology of the cell, they were very quick to exploit the techniques and successes of these new investigations. One of the biggest questions left from the 1950s was the nature of variation within populations. Is enormous genetic variation always present in a population, ready to be used as raw material for adaptation when the environment changes? Do various mechanisms, such as selection for rareness, sustain variation in populations? Or does selection purify populations almost immediately, either eliminating new mutations at once or rapidly fixing them within the group? Clearly this is no small matter, because if lots of variation were constantly available, then selection as the chief cause of change would be much more plausible. In that case, selection would be a flexible and dynamic process, as genomes respond quickly to threats or take advantage of opportunities. Species would not have to wait patiently for the right mutation to arrive randomly. If one gene could not do the job, a veritable library of alternatives could be taken up and used.

Molecular biology gave evolutionists the tools to cut through this thicket of questions. Dobzhansky's student Richard Lewontin (and others) developed the technique of gel electrophoresis, which exploited the different electrostatic charges carried on different biological macromolecules.[2] It revealed at once the molecular genetic differences between specimens, and the nature of these differences. Lewontin and his associates were able to show that a truly massive amount of variation exists within natural populations—in humans, mice, fruit flies, plants, wherever they looked.

Variation in populations was an issue that much concerned the evolutionists of the 1940s and 1950s. It was natural therefore that it would be taken up when molecular biological techniques became available. What is truly exciting is the way that other areas of evolutionary interest—areas that were neglected by Dobzhansky and his coworkers—have been reclaimed and turned into fertile fields of research.

Charles Darwin always thought of embryology as one of the really important parts of his theory, and he took great pride in the way that evolution explained the similarities among embryos of different species. Natural selection in particular threw light on the nature of embryological development, in Darwin's view. At the turn of the nineteenth century, during evolution's long non-Darwinian phase, embryology continued to absorb the energies of evolutionists, although (as with everything else) phylogeny—the change of species into other species over time—was their main focus. In the 1940s, when the synthetic theory (as it was called in America; in England, it was better known as neo-Darwinism) came into being, precisely because embryology was a major part of what the synthesis's architects were rejecting, the development of individual organisms from embryos to adults was ignored. In recent years, this has changed completely, and evolutionary development (evo-devo) has become the hottest part of the discipline. Much molecular-informed activity is being directed toward an understanding of development and the ways in which it can affect the course of evolution.[3]

The most dramatic discoveries in evo-devo have been quite unexpected DNA homologies. It turns out that organisms as different as fruit flies and humans share considerable amounts of practically unaltered DNA, especially those stretches that are involved in development itself—ordering the rates and ways in which the parts of the body are formed (heads before legs and so forth). The jury is still out on the precise significance of all of this. Some seem to think that selection will now have to take a back seat in evolution: "The homologies of process within morphogenetic fields provide some of the best evidence for evolution—just as skeletal and organ homologies did earlier. Thus, the evidence for evolution is better than ever. The role of natural selection in evolution, however, is seen to play less an important role. It is merely a filter for unsuccessful morphologies generated by development. Population genetics is destined to change if it is not to become as irrelevant to evolution as Newtonian mechanics is to contemporary physics."[4]

Others find these new discoveries less threatening to the established order. Darwin himself would probably not have been particularly fazed. In his little book on orchids, written just after the *Origin*, he stressed how organisms use and reuse different body parts, and how things change only with good reason. Hence, the molecular homologies—like all cases of phylogenetic inertia—might simply reflect the fact that they do the job well in all organisms, fruit flies and humans. By analogy, "the fact that all tires are round more likely means that round wheels are optimally functional than that tire companies are somehow constrained by the round shape of their existing molds. Thus phylogenetic inertia is not an alternative to natural selection as a mechanism of persistence, and evidence of the former is not evidence against the latter."[5]

The Evolution of Social Behavior

The one area of modern evolutionary biology that is unambiguously adaptationist is the study of animal social behavior, a topic to which Darwin himself gave considerable attention in the *Origin*. The major breakthroughs came in the 1960s with a series of models showing that social behavior could be explained by natural selection. The most notable work came from the Englishman W. D. Hamilton, who solved a major problem that had been particularly puzzling to Darwin. The author of the *Origin* was convinced that selection rewards adaptation in individuals, not in groups as a whole. So why do most female ants, bees, and wasps devote their lives to the good of the nest and forgo reproduction themselves?

Hamilton cracked this mystery among the social insects by invoking what came to be known as kin selection—reproduction by proxy, as it were. The behavior of aiding a fellow species member can be selected and perfected in situations where the individuals doing the aiding are closely related to those that benefit. As the recipients successfully reproduce, the aiders pass on copies of many of their own genes. In most animal species, individuals prefer to do their own reproducing, because that maximizes the number of their genes that go into the next generation. But Hamilton pointed out that the social insects are a special case.

In most sexual organisms, like humans, cells contain two half sets of chromosomes (the carriers of the genes), one half coming from the father and the other from the mother. So any parent and any child are related 50 percent. Because any half set of genes is drawn from the parent's whole set, and because these genes are randomly shuffled, the half sets siblings receive will not be identical but themselves only 50 percent alike on average. Since any two siblings receive 25 percent identical genes from their father and 25 percent identical genes from their mother, any two siblings are as closely related (50 percent) as are any parent-child combination (50 percent). In the hymenoptera, however, things are different. Although females have both a mother and a father, and hence two half sets of chromosomes, males are produced from unfertilized eggs. They therefore have only one half set, which comes entirely from their mothers—and they have only this one half set to pass on. This means that sisters who share the same father are already 50 percent related on their dad's side; the mother's contribution adds an additional 25 percent. This means that sisters are more closely related (75 percent) than mothers and daughters (50 percent). So from an evolutionary perspective a female nest member is better off raising a fertile sister (the queen), with whom she shares 75 percent of her genes, than fertile daughters, with whom she would share only 50 percent. From an individual-selection perspective, sociality pays.[6]

In line with Hamilton's insights, other proposals made full use of mathematical ideas borrowed from game theory. Serious attention was given to McEwan's hero's problem, namely, how do you balance the need to cooperate—a vital adaptation for a social species—with the urge to serve oneself, for which the popular science writer Richard Dawkins has coined the memorable term the "selfish gene."[7] Particularly important was the idea of an "evolutionarily stable strategy," where in a group of organisms different members do different things (or have different features), with ratios balanced because if one kind grows or declines, the advantages for others are affected. We have already encountered such a strategy in the Darwin/Fisher explanation of equal sex ratios.

After the theoreticians came the empiricists, and they showed just how powerful the new evolutionary understanding of social behavior could be. Light was thrown all across the animal kingdom. Why do male lions work in pairs or triplets rather than singly? Because members of male bands tend to be brothers, and kin selection comes into play. Why do chimpanzees have huge testicles and copulate a lot, whereas gorillas, which are much larger and stronger than chimps, have small testicles and copulate infrequently? A male gorilla fights with other males for the privilege of owning a harem; once he chases off his competitors, the alpha male enjoys his leisurely pick of females in the harem, with no interference. But a male chimp, being a tree climber, cannot grow sufficiently large to defend a harem, and so he lives in a large group with members of both sexes. With so much daily competition for copulations and reproductive success, a male chimp needs a pair of large, productive testicles. Human males, who tend to strike a balance between competition for copulations and faithfulness to partners, are somewhere in the middle between gorillas and chimpanzees when it comes to testicle size.

And that's not all. Why does a male dung fly give up a female before all her eggs are fertilized and go off searching for another female? Because it is better to look for fresh eggs than to scrape the bottom of the barrel looking for old ones. Why are hedge sparrows into group sex, and who looks after the babies? Their habitats make monogamy difficult to defend, and the males who get the most copulations help most at the nest. Why are the eggs that the cuckoo leaves in a robin's nest camouflaged to look like robin eggs, while the eggs it leaves in the hedge sparrow's nest have no such camouflage? The hedge sparrows, which have only been recently parasitized by cuckoos, do not yet have counter adaptations in place that would allow them to destroy the offending egg, even if it looks nothing like their own.

Why do healthy and dominant female mammals tend to have male offspring, and less healthy and dominant females tend to have female offspring? Because of the asymmetry between male and female reproductive chances. Theoretically, any one male can have many offspring. Theoretically, any one female can have a much smaller number. Practically, since males compete against other males to sire the small number of offspring that females can produce, some males can and will have many offspring by many females, and some males will have few or none. Females, on the other hand, can and generally will reach their biological quota if they are healthy. Success, for males, depends on many things, but health and parental status are crucial factors. Males born to higher-status healthy females are more likely to succeed than males born to lower-status less healthy females, and so nature has taken a hand in bringing about the best reproductive chances for both kinds of female. Apparently something analogous to this natural selection process can happen within human societies also. Some members of higher castes in India kill female infants and marry their sons to females of lower castes. Conversely, some lower status groups (Hungarian gypsies, for instance) produce suspiciously few male offspring.[8]

Again and again in this newly invigorated field of the evolution of social behavior—also called sociobiology—an adaptive approach has proved fruitful. Selection is the key to understanding, and this was the triumphant message of the masterwork that surveyed the field and brought it all together beneath one inclusive Darwinian hypothesis of selective advantage: *Sociobiology: The New Synthesis,* published in 1975 by the Harvard-based student of insect social behavior, Edward O. Wilson. From slime molds to humans, it was one all-inclusive picture.

Paleontology

To complete this brief survey of modern evolutionary biology, we turn now to the topic that most lay people think of first when evolution is mentioned: paleontology, with its fossils (and fantasies) of exotic extinct species. The fossil discoveries of the past half century have been absolutely stupendous. Combined with new and more powerful techniques of classification—techniques that often rely heavily on the use of computers to crunch large amounts of information—we now have a much clearer picture of life's history than before. For instance, the evidence is very strong that the dinosaurs did not all become extinct. They still exist, except that now we call them birds. This question of the extinction of the dinosaurs was still highly controversial in Thomas Henry Huxley's time, but today there is consensus.

The fossil record of humans is also much richer than it used to be. Best known is Lucy, the full skeleton of a specimen of *Australopithecus afarensis,* who lived about four million years ago, walked upright (although not as well as us), and yet had an ape-sized brain. She is only one of many specimens of her own and other related species. Incredibly exciting is the recent discovery of a new species, *Homo floresiensis*, a small hobbitlike creature whose remains have been found on one of the islands of Indonesia. Apparently no more than 20,000 years ago—and possibly more recently—this creature within our own genus but of a different species still existed. It was no more than three feet tall and had a chimpanzee-sized brain, but it was about as close to human as it is possible to be without actually being one of us.

Molecular comparisons suggest that we humans are more closely related to chimpanzees than chimpanzees are to other apes, including gorillas. Precisely when we split from the ape line is still a matter of debate, but six million years ago is a close approximation—a much shorter time than people used to think and but a mere blink in the nearly four-billion-year history of life on earth.

There have been exciting theoretical developments also. The late Stephen Jay Gould, along with his fellow paleontologist Niles Eldredge, has famously argued that the course of evolution is jerky rather than smooth.[9] The Eldredge/Gould hypothesis of punctuated equilibria is still very controversial, and not the least of the controversy is whether, if true, it calls for causal mechanisms other than selection. Gould himself was confusing on this issue, sometimes inclining one way and sometimes another. Now, as in the days of Marsh and Cope, paleontology provides some of the best evidence for the fact of evolution, but it is slow to confirm the significance of selection as evolution's prime mover.

Still, no modern paleontologist would deny that selection is a very powerful tool for casting light on a massive number of problems raised by the fossil record. One of the most interesting is the long debate about those diamond-shaped plates that run down the back of stegosaurus. Were they for attack or defense? Were they for sexual attraction, like the tail of the peacock, and otherwise a burden on the brute, or were they for something else yet again?

> When Stego, out foraging, tarried
> To serve up the herbs near his track,
> He felt well equipped, for he carried
> Two rows of bone plates on his back.

More prosaically, recent thinking is that the plates may have serviced their dinosaur owner by regulating body temperature, and much effort has been devoted to showing that the plates have the same characteristics that engineers design into blades used for heat transfer in electrical cooling stations—a paradigmatic example of selection for functionality.

Not that all of the evolutionist's problems have been solved, or are even close to being solved. The problem of the origin of life has always been a major headache for evolutionists. In order to tackle the problem adequately, one needs training that evolutionists generally lack—a huge amount of biochemical knowledge, among other things. Still, some progress has been made on the problem. It now looks as though ribonucleic acid (RNA), rather than its sister molecule DNA, was a key component in the origin of life, because of its unique ability to self-replicate. Researchers have already succeeded in coaxing molecules of RNA to grow and proliferate spontaneously in the laboratory. The problem is not that investigators have no ideas about how life got started but that they have too many ideas, and much work is needed to winnow the less plausible and fruitful ones. Understanding of how a functioning cell might have formed, once life got going, has also blossomed in recent years.

This parade of successes is not intended to deny the large gaps in our knowledge of evolutionary mechanisms that remain. There is plenty of unfinished business to keep researchers working away, productively and happily, for many years. A functioning discipline of evolutionary biology—in large part, if not exclusively, Darwinian—is alive and actively churning out new questions as well as answers to the age-old mysteries of life.

Still a Secular Religion?

Let us move on from the professional to the popular side of evolutionary biology. Does a popular side still exist, and if it does, is it still proper to speak of it as a secular religion? The first part of this question is easy to answer. A popular side does exist. Evolution enjoys as high a profile today among lay people as it ever has. Newspaper articles, television programs, magazine discussions, and a flow of attractive books keep evolution and evolutionists—their triumphs, disagreements, and frustrations—constantly in the public eye.

Both professional science writers and fully professional evolutionists—one thinks particularly of Dawkins in England and Wilson and Gould in America—have been very successful in talking to nonscientists about the science of evolutionary biology and its implications. And in the humanities, a full-scale industry has grown up which studies evolutionary biology from various disciplinary perspectives—philosophy, history, sociology, rhetoric, and other fields.[10] Some of the works of these humanistic scholars can be read only by their colleagues (if that), but others—for instance, recent biographies of Charles Darwin—are very accessible to general readers.

But if, undeniably, much of popular evolutionary biology is simple and straightforward science writing, and not necessarily an attempt to send a religious or metaphysical message, does that mean that evolution as a "secular religion" is dead? Obviously, we are not talking about robes, rituals, and a priesthood but of evolution versus evolution*ism*. As we would expect, academic evolutionists deny any religious associations in their field—after all, they are scientists who have only recently dragged themselves up to full professional status, and would just as soon forget evolution's checkered past. Richard Dawkins is characteristically eloquent on this topic:

> It is fashionable to wax apocalyptic about the threat to humanity posed by the AIDS virus, "mad cow" disease, and many others, but I think a case can be made that *faith* is one of the world's great evils, comparable to the smallpox virus but harder to eradicate. Faith, being belief that isn't based on evidence, is the principal vice of any religion . . . Given the dangers of faith—and considering the accomplishments of reason and observation in the activity called science—I find it ironic that, whenever I lecture publicly, there always seems to be someone who comes forward and says, "Of course, your science is just a religion like ours. Fundamentally, science just comes down to faith, doesn't it?" Well, science is not religion and it doesn't just come down to faith. Although it has many of religion's virtues, it has none of its vices. Science is based upon verifiable evidence. Religious faith not only lacks evidence, its independence from evidence is its pride and joy, shouted from the rooftops.[11]

Things were not quite this simple in the years after Darwin, and it is instructive to ask if they are indeed this simple today. Let us frame our question about evolutionism as a secular religion in two ways, as a negative and a positive. First, is evolutionary thinking used to combat or deny other religions, specifically the validity of some or all forms of Christianity? Second, is evolutionary thinking used to promote a world picture with norms of its own, which may or may not coincide with those of the religious but which are proposed (and justified) essentially on evolutionary grounds alone? To extend this last question to the extreme, does evolutionism entail its own brand of postmillennial theology?

We cannot make a universal case. The claim that every evolutionist is some kind of secular religious convert is demonstrably false. Many of today's younger professionals have no special interest in culture and values, and certainly no interest in using their science as a vehicle to express and promote such values. Indeed, for some, science is a way of getting *away* from values. The highly respected English sociobiologist Geoffrey Parker put it this way: "It's a liberation really. You can choose to be highly superstitious, to believe in all sorts of things if you wish to. Some people erect huge superstructures of superstition—they can't allow themselves to go out on Friday the thirteenth. They must do this before they go to bed. Whatever. Salt over the shoulder. Mustn't wear green. And it goes on and on. I really think in a way I do see science—being a scientist perhaps—as some sort of liberation from those things."[12]

Recognizing therefore the restricted nature of our inquiry, we can boil down the first question to this: whether evolutionists argue that their position wipes out the possibility of any reasonable religious commitment—that you must choose between God or Darwin. There are many who argue that one must make this choice. The leading Chicago evolutionary biologist, Jerry Coyne (a student of Lewontin), speaks for this crowd:

> Christians have good reason to feel uncomfortable about Darwinism. The fossil record shows that the Genesis version of creation is manifestly wrong if read literally, and one is left either questioning the authority of the Bible or recognising that it is a prolonged exercise in metaphor—and as such open to endless

interpretation. Moreover, it is difficult for a committed Darwinist to view humans, who form one side branch of the primate lineage, as the principal object of creation. For many biologists, the knowledge that *Homo sapiens* is only one of many evolved species—albeit one with a large brain and an extensive culture—makes it difficult to find any preordained meaning or purpose in human existence. Finally, if one applies the same empirical standards to Christianity as scientists do to Darwinism, religion suffers: we have far more evidence for the existence of dinosaurs than for the divinity of Christ.[13]

In reviewing a book of mine that answered positively the question posed in its title, *Can a Darwinian Be a Christian?* Coyne wittily quoted George Orwell. "One has to belong to the intelligentsia to believe things like that. No ordinary man could be such a fool."

One gets more than just disagreement on this question. Sometimes one gets outright hostility. Dawkins, writing about the Pope, sounds as though he was born and bred in a Protestant family from Northern Ireland (which he wasn't): "Given a choice between honest to goodness fundamentalism on the one hand, and the obscurantist, disingenuous doublethink of the Roman Catholic Church on the other, I know which I prefer." But having characterized his own move to atheism from religious belief as a "road to Damascus" experience, Dawkins is ecumenical in his hostility toward Christianity. Explicitly, he uses terms that we associate with the hatred of one religion by subscribers to another.

> I am considered by some to be a zealot. This comes partly from
> a passionate revulsion against fatuous religious prejudices, which
> I think lead to evil. As far as being a scientist is concerned, my
> zealotry comes from a deep concern for the truth. I'm extremely

hostile towards any sort of obscurantism, pretension. If I think somebody's a fake, if somebody isn't genuinely concerned about what actually is true but is instead doing something for some other motive, if somebody is trying to appear like an intellectual, or trying to appear more profound than he is, or more mysterious than he is, I'm very hostile to that. There's a certain amount of that in religion. The universe is a difficult enough place to understand already without introducing additional mystical mysteriousness that's not actually there. Another point is esthetic: the universe is genuinely mysterious, grand, beautiful, awe inspiring. The kinds of views of the universe which religious people have traditionally embraced have been puny, pathetic, and measly in comparison to the way the universe actually is. The universe presented by organized religions is a poky little medieval universe, and extremely limited.

I'm a Darwinist because I believe the only alternatives are Lamarckism or God, neither of which does the job as an explanatory principle. Life in the universe is either Darwinian or something else not yet thought of.[14]

In a tradition that goes back to Darwin himself, Dawkins identifies a major element of his nonbelief as being the pain and suffering caused by natural selection. He quotes from a letter written by Darwin to Asa Gray: "I cannot persuade myself that a beneficent & omnipotent God would have designedly created the Ichneumonidae with the express intention of their feeding within the living bodies of caterpillars, or that a cat should play with mice."[15]

Some evolutionists are less overtly hostile to religion than Dawkins. Stephen Jay Gould spoke of science and religion as different worlds of experience and understanding—what he called different "Magisteria." Supposedly, they both have their own roles to play, and claims by one cannot be corroded by the claims of the other. But as one looks closer, it turns out that Gould's match is uneven. Religion apparently can tell us only about morality—about how we ought to behave. Any claims about existence or origins that it might make are almost certainly false. So much for Creation, Incarnation, Resurrection, the Second Coming, and so forth. And any thought that we might be the favored children of God, that we might be the (or a significant) reason for creation, is simply hubris. Like Coyne, Gould was a major booster of the contingency of life: "Since dinosaurs were not moving toward markedly larger brains, and since such a prospect may lie outside the capabilities of reptilian design . . . we must assume that consciousness would not have evolved on our planet if a cosmic catastrophe had not claimed the dinosaurs as victims. In an entirely literal sense, we owe our existence, as large and reasoning mammals, to our lucky stars."[16]

Finally, we have Edward O. Wilson. He is quite positive about religion. Human nature needs religion to function, he has stated. Religions are biologically adaptive. "Above all they congeal identity. In the midst of the chaotic and potentially disorienting experiences each person undergoes daily, religion classifies him, provides him with unquestioned membership in a group claiming great powers, and by this means gives him a driving purpose in life compatible with his self-interest." The conclusion Wilson draws, however, is not that this speaks to the validity of the world's traditional religions but rather to their replacement by an evolutionary picture.

> But make no mistake about the power of scientific materialism. It presents the human mind with an alternative mythology that until now has always, point for point in zones of conflict, defeated traditional religion. Its narrative form is the epic: the evolution of the universe from the big bang of fifteen billion years

ago through the origin of the elements and celestial bodies to the beginnings of life on earth. The evolutionary epic is mythology in the sense that the laws it adduces here and now are believed but can never be definitively proved to form a cause-and-effect continuum from physics to the social sciences, from this world to all other worlds in the visible universe, and backward through time to the beginning of the universe. Every part of existence is considered to be obedient to physical laws requiring no external control. The scientist's devotion to parsimony in explanation excludes the divine spirit and other extraneous agents. Most importantly, we have come to the crucial stage in the history of biology when religion itself is subject to the explanations of the natural sciences. As I have tried to show, sociobiology can account for the very origin of mythology by the principle of natural selection acting on the genetically evolving material structure of the human brain.

If this interpretation is correct, the final decisive edge enjoyed by scientific naturalism will come from its capacity to explain traditional religion, its chief competition, as a wholly material phenomenon. Theology is not likely to survive as an independent intellectual discipline.[17]

And this brings us to our second question. Is evolution used by some of its proponents to urge us into action? Is it promoted as a guide to and justification for morality? Again, the question is not whether evolutionists happen to make moral prescriptions, but whether they claim that these moral prescriptions are mandated by the principles of biological evolution. It may come as a surprise that the arch atheist among evolutionists, Richard Dawkins, is very uncomfortable with the notion that moral prescriptions can be derived from the course of evolution. Although (as we shall see in the next chapter) he is ardent for biological progress, he is severe in his criticism of Prince Charles, heir to the British throne, for using an appeal to nature and evolution in mounting his argument opposing the use of genetically modified foods. Dawkins echoes Thomas Henry Huxley that what has been produced by evolution is not necessarily good. In the case of foodstuffs, Dawkins thinks that modification by humans was the key to improvement, and not a barrier to "natural" improvement through evolutionary means.[18]

But criticisms of royalty aside, Dawkins's overall stand on evolution, with its supposed negative implications for the existence of the Christian deity, has significant implications for his thinking about social and moral issues. "The human psyche has two great sicknesses," he writes, "the urge to carry vendetta across generations, and the tendency to fasten group labels on people rather than see them as individuals. Abrahamic religion mixes explosively with (and gives strong sanction to) both. Only the willfully blind could fail to implicate the divisive force in religion in most, if not all, of the violent enmities of the world today. Those of us who have for years politely concealed our contempt for the dangerous collective illusion of religion need to stand up and speak out."[19] Religion leads to evil. Darwinism counters religion. Ergo, Darwinism counters evil.

In the face of such strong emotions, who can deny that evolution functions as more than just a scientific theory in Richard Dawkins's worldview? "All the great religions have a place for awe, for ecstatic transport at the wonder and beauty of creation. And it's exactly this feeling of spine-shivering, breath-catching awe—almost worship—this flooding of the chest with ecstatic wonder, that modern science can provide. And it does so beyond the wildest dreams of saints and mystics."[20] For Dawkins, contemplation of the natural world through the eyes of science is a religious experience.

Others, no less passionate, take a more traditional approach to these matters. Although they would never use the term "social Darwinian," their philosophy of life is grounded in the belief that natural selection should be left alone to do its work, however much that may entail suffering. Biological progress is the key to this worldview. W. D. Hamilton, for example, idolized the memory of Fisher and shared his progressive vision of evolution as moving slowly upward, with humans as the pinnacle and with civilization as the greatest threat to human progress, because it thwarts the ability of natural selection to weed out weak variants. Indeed, for Hamilton we are already on a path downward and must take steps to reverse the fall.

> I predict that in two generations the damage being done to the human genome by the ante- and postnatal life-saving efforts of modern medicine will be obvious to all and be a big talking point of science and politics . . . Metastability—the possibility of non-self-limiting breakdown—will hold with respect to potential epidemic disease agents, to potential physical disruptions (sudden climate change, asteroid impact, and the like), and with respect to sheer physical inabilities in daily life. The servicing of an essentially machine- and electronics-based culture will be ever more liable to an escalating breakdown. After 80 years I predict people will have begun to think seriously about these points. By then, I admit, there will still be enough time to back out, although it will only be to an accompaniment of considerable suffering.[21]

In Hamilton's view, we need to toughen up significantly and make sure that the sick and handicapped do not survive—or if they do survive, that they do not reproduce.

Crossing back to America, we have the position of Wilson, who is an ardent progressionist: "In my case, I have grown up in a culture that is heavily devoted to millenarianism, that is the belief that—I don't believe this—that Christ will soon come and we'll all go to Paradise. Boosterism—the American spirit of entrepreneurship and the belief in the unlimited ability of individuals to rise and prosper. The Protestant quality of belief in the work ethic and the rewards of work." This can-do environment seems to have imbued Wilson with a strong belief in progress, minus the Christian dogma. This is reflected in the world of organisms: "The overall average across the history of life has moved from the simple and few to the more complex and numerous. During the past billion years, animals as a whole evolved upward in body size, feeding and defensive techniques, brain and behavioral complexity, social organization, and precision of environmental control—in each case farther from the nonliving state than their simpler antecedents did." He adds: "Progress, then, is a property of the evolution of life as a whole by almost any conceivable intuitive standard, including the acquisition of goals and intentions in the behavior of animals."[22]

Like Hamilton, Wilson does not see this progress as morally neutral. Humans rise above hippopotami, and hippopotami above horseflies. Wilson is explicit in believing that from biological progress comes value, and that humans are therefore the most valuable, the most worthy of cherishing. Appropriately, Herbert Spencer is his personal hero. And like all such evolutionary ethicists, Wilson thinks that action is needed to reverse downward trends and to ensure ongoing progress. Through and only through biology can we find answers to the meaning of life. For Wilson, a deeply committed naturalist, the key to human survival—the loss of which is the biggest threat that we face—is biodiversity, the tangled bank of which Charles Darwin wrote at the end of the *Origin of Species.* Wilson draws attention to the ways in which human culture impinges on the natural environment, to the detriment of all.

At the end of his masterwork, *Sociobiology: The New Synthesis*, sounding much like those old-time preachers of his childhood, Wilson calls upon us to change our ways. We have but a few years left before it will be too late, and the rain forests will be gone for ever. Repent! Repent! Wilson has repeated this theme many times in subsequent years. Human nature is "biophilic," we have a genetically innate love of nature; a world of plastic would be deadening, literally as well as metaphorically. This, for Wilson, is an evolutionary fact, not a cultural artifact. In *The Future of Life*, Wilson writes that "a sense of genetic unity, kinship, and deep history are among the values that bond us to the living environment. They are survival mechanisms for ourselves and our species. To conserve biological diversity is an investment in immortality."[23]

But why, ultimately, should we care about the survival of humans, except as a matter of self-interest? Why should it be a moral question whether we survive or not? Why should human immortality matter? Wilson's answer is loud and clear: because evolution is progressive and we have come out on top. Not to preserve humans would be to reduce value, and that is wrong.

Wilson's type of thinking has not gone unchallenged. Despite an early flirtation with progressionism—think of the Spencerian undertones of the term "punctuated equilibria"—Stephen Jay Gould was, for many years, a strong and vocal critic of all suggestions that the course of nature leads upward to humankind. He spoke of the idea as "a noxious, culturally embedded, untestable, nonoperational, intractable idea that must be replaced if we wish to understand the patterns of history." Biological progress is a delusion engendered by our refusal to accept our insignificance when faced with the immensity of time. Gould's point was not that complexity, in some sense or another, has not increased. A human is more complex than a single-celled organism, on any scale or definition of complexity that can be imagined.

The point rather is that this rise in complexity has no underlying meaning, and that nothing in the theory or fact of evolution drives its realization.[24]

The appropriate analogy is with a drunken man, staggering along a sidewalk. On one side is a wall, and on the other side is a gutter. He cannot go through the wall, and his staggering path will eventually land him in the gutter. Likewise with life. You can't get less complex than simple (simple is the wall), but you can get more complex. It is simply a matter of chance operating over long stretches of time. It is not a matter of meaning. Complexity—that is, biological progress— is neither causally produced nor better than simplicity. But it is more likely.

Biological progress per se is not the real issue here, however. The real issue is whether some evolutionists today use the supposed progressiveness of evolutionary theory to promote social and ethical programs. And indeed they do. Exceptions like Gould often tend to prove the rule. Gould's sudden switch from accepting and arguing for biological progress to denying and arguing against it came about because he grew, increasingly, to believe that doctrines of biological progressionism are antithetical to certain notions of social progress that he endorsed wholeheartedly. Gould saw (with some justification) that biological progressionism is too readily a vehicle for claims about the superiority of one human group over another—English over the Irish, gentiles over Jews, whites over blacks—and in order to combat this racism, Gould spent over two decades fighting tooth and nail against all genuine forms of progressionism. His real objection was moral or metaphysical, not scientific. In this sense, he was using the idea of evolution to promote an optimistic eschatology as much as any straightforward progressionist.

Gould was a controversial figure, and debate continues about the ultimate worth of his scientific contributions to biology, let alone the philosophical juice he extracted from them. But putting him to one side, many others reinforce the point. After Darwin, the claim could no longer be made that (absolute) biological progress is value-neutral, something that one can simply read from the evolutionary picture, justified by theory. Rather, progress was a value that humans added. And in the adding, they shifted from a scientific theory of evolution to a quasi-religious commitment to evolutionism. Those who made this shift were not fringe figures, with no standing or respect in the evolutionary profession. To the contrary, there are no more honored names in the field than William D. Hamilton and Edward O. Wilson. The outstanding leaders of the discipline are among those people who promote social and ethical programs on the basis of their evolutionary commitment.

This holds true in England and even more so in America. In this sense, evolutionary biology—Darwinian evolutionary biology—continues to function as a kind of secular religion. It offers a story of origins. It provides a privileged place at the top for humans. It exhorts humans to action, on the basis of evolutionary principles. It opposes other solutions to questions of social behavior and morality. And it points to a brighter future if all is done as it should be done, in accordance with evolutionary theory. Wilson may be right that he has shucked the literal apocalyptic commitments of his childhood, but if he is not committed to a postmillennial theology, I do not know who is.

II

Nature as Promise

Tell me not here, it needs not saying,
    What tune the enchantress plays
In aftermaths of soft September
    Or under blanching mays,
For she and I were long acquainted
    And I knew all her ways.

On russet floors, by waters idle,
    The pine lets fall its cone;
The cuckoo shouts all day at nothing
    In leafy dells alone;
And traveller's joy beguiles in autumn
    Hearts that have lost their own.

On acres of the seeded grasses
    The changing burnish heaves;
Or marshaled under moons of harvest
    Stand still all night the sheaves
Or beeches strip in storms for winter
    And stain the wind with leaves.

*T*hese lines from the English poet A. E. Housman seem to describe Constable country—soft September, russet floors, falling pine cones. But the final verse has a very different, more somber message.

For nature, heartless, witless nature,
    Will neither care nor know
What stranger's feet will find the meadow

And trespass there and go,
Nor ask amid the dews of morning
If they were mine or no.

The first two of these last lines were picked out by Richard Dawkins as representing his conclusion from Darwinian evolution: "In a universe of blind physical forces and genetic replication, some people are going to get hurt, other people are going to get lucky, and you won't find any rhyme or reason in it, nor any justice. The universe we observe has precisely the properties we should expect if there is, at bottom, no design, no purpose, no evil and no good, nothing but blind, pitiless indifference. As that unhappy poet A. E. Housman put it: 'For Nature, heartless, witless Nature / Will neither know nor care.' DNA neither knows nor cares. DNA just is. And we dance to its music."[1]

We cannot blame (if that is the right word) either Darwin or Dawkins for this harsh, stark vision. Matthew Arnold's "Dover Beach" came before either.

. . . the world, which seems
To lie before us like a land of dreams,
So various, so beautiful, so new,
Hath really neither joy, nor love, nor light,
Nor certitude, nor peace, nor help for pain . . .

But many fear that Darwinism—too ably aided by its acolyte, Dawkins—makes the vision a real force and menace. Can Christian faith be maintained in the face of such a threat?

Standing on the threshold of the twentieth century, one might have predicted that, theologically speaking, this would become the age of evolution. That God's having created the world in a slow sequential fashion, through natural law, would have been the central focus of any serious attempt to understand Christianity and its meaning for humanity. With hindsight, we can see that our prediction would have been false. While evolution has become the standard belief of many Christians, Catholics as well as Protestants, one would be mistaken to think that evolution conquered all. From Huxley to Dawkins, many biologists scorned and denied Christianity—and this caused religious people to pull back and turn away to other more friendly explanations for the world they inhabited.

This move was greased by the belief among many twentieth-century Christians that progress is a false philosophy. After the experience of two dreadful world wars, a worldwide depression, horrendous dictatorships in Germany, Russia, Italy, and Spain (not to mention countless repressive regimes in non-European nations), and much more, many Christians came to agree with Aldous Huxley that progress—and anything connected to it, including evolution—was cruel heresy. As the high-Anglican poet T. S. Eliot wrote in 1948: "We can assert with some confidence that our own period is one of decline; that the standards of culture are lower than they were fifty years ago; and that the evidences of this decline are visible in every department of human activity."[2]

Another major reason for the rift between evolution and Christianity was a theological development. Echoing earlier thinkers, the Swiss theologian Karl Barth argued strongly that reason can give no true understanding of God. Faith alone can take us into a proper relationship with the Creator.[3] Faith is mediated through Jesus Christ, and hence scripture is the only foundation of proper religion. The fruits of reason and observation—that is, science—count for little. For Barth and his followers, natural theology, old or new, pre- or post-evolutionary, was simply a false and misleading approach to the divinity.

In Barth's case, this theology grew out of an attempt to sever once and for all the link between German Christianity and movements such as National Socialism in which reason gave false grounds for the significance of the state and its rulers. Not everyone shared these particular worries, and natural theology itself showed considerable resilience in the twentieth century. Many agreed, however, that evolution's dark subtext could be used to support doctrines of racial superiority and programs of genetic improvement. Obviously, one can be an evolutionist and scorn eugenics, but evolution and eugenics have not been strangers, historically. And for many Christians, this past association was reason enough to avoid all biological hypotheses.

Yet, a minority of Christian thinkers wrestled with evolution in an attempt to use it fruitfully in building an adequate theology for the modern world. These people tended to be scientists as well as believers—thinkers who realized the power and importance of Darwin's legacy and tried to see how Christians could make sense both of a living world produced by descent with modification and of Jesus Christ as the redeemer of humankind. They tended also to be at one with those who took an optimistic view of history and human progress. Their urge was to join nineteenth-century thinkers in seeing evolution as thrusting ever upward and culminating in the creation of our species.

The most important of these thinkers was neither British nor American nor even a Darwinian when it came to causes. He was a French Jesuit paleontologist, Father Pierre Teilhard de Chardin. In *The Phenomenon of Man*—published in 1955, after his death, because his superiors were frightened of its content—he saw an upward evolution through the realm of life (the biosphere), to the realm of humans and consciousness (the noösphere), and then even further onward and upward to the Omega Point, which in some way he identified with the Godhead, Jesus Christ. "An ever-ascending curve, the points of transformation of which are never repeated; a constantly rising tide below the rhythmic tides of the ages—it is on this essential curve, it is in relation to this advancing level of the waters, that the phenomenon of life, as I see things, must be situated."[4]

This upward movement is marked by critical stages of transformation from one form to another. "In every domain, when anything exceeds a certain measurement, it suddenly changes its aspect, condition, or nature. The curve doubles back, the surface contracts to a point, the solid disintegrates, the liquid boils, the germ cell divides, intuition suddenly bursts on the piled-up facts. Critical points have been reached, rungs on the ladder involving a change of state." The key transformation is the move to human consciousness. "The greatest revelation open to science today is to perceive that everything precious, active and progressive originally contained in the cosmic fragment from which our world emerged is now concentrated in and crowned by the noosphere."[5]

And so we may look to future triumphs, the end point of evolution being the "superior stage which humanity appears destined to achieve if it succeeds in totalizing itself completely upon itself, body and soul, by pushing to the end the movement of which it is the historical culmination." There are intimations here of the group-consciousness notion that excited Stapledon and Wright: "If socialization is nothing but a higher effect of 'body formation'—*corpuscularisation*—then the noosphere as the last and highest product of this process, can get full and decisive meaning only under *one* condition: if it is considered, to its full extent, as one single infinite body in which, after more than six hundred million years, the work of forming the brain—*cerebration*—is completed in the noosphere."[6]

Because of his troubles with the Catholic Church, instead of admitting frankly that he was spinning metaphysical visions, Teilhard presented his own thinking as pure science. The gambit was unfortunate, for where he approached empirical questions, especially those to do with underlying causes, he showed himself out of tune with modern science. Not surprisingly, Teilhard had been much influenced by the philosophy prominent in his youth, particularly the vitalism of his fellow Frenchman Henri Bergson, who argued that life forces, *élans vitaux*, control living things and somehow lead to their striving ever higher and onward. To this vitalistic backbone Teilhard added various now-discredited secondary mechanisms, such as the Lamarckian inheritance of characteristics.

The inevitable consequence was that many conventional scientists were scathing in their criticisms of Teilhard. Peter Medawar, a Nobel Prize–winning immunologist, wrote a review that could have come from the poisoned pen of Thomas Henry Huxley. Stephen Jay Gould played the cruelest trick, accusing Teilhard (on the flimsiest evidence) of having perpetrated the Piltdown Hoax—the fabrication in the early twentieth century of a man-ape skull—and therefore of being unworthy of serious regard as a scientist. Yet, Teilhard also won many followers. The nonbeliever Julian Huxley responded very positively, as did the Russian Orthodox Christian Theodosius Dobzhansky. Even if his influence was not direct, Teilhard inspired many in the second half of the last century to try again to see if evolutionary thinking could be incorporated into a Christian vision of the world—some way of showing how God created in an evolutionary fashion, guaranteeing the arrival of humans, and how this perspective puts the emphasis on the future as much as on the past—and how we must think in terms of God's intended plans as much as his completed actions.[7]

Can Selection Lead to Progress?

Is there anything in contemporary Darwinism that would give one hope for such aspirations? Leaving now to one side those evolutionary biologists who insist that selection alone cannot be the full causal story for life, let alone for progress, and also moving away from vitalism, can we find in selection itself the progressive push that would have been needed to produce humans? Although everyone respects Fisher as a scientist, his metaphysics has evoked less enthusiasm. Apart from his associations with eugenics, the progressiveness of his fundamental theorem seems a bit too much like putting in precisely what you then want to pull out.

More attractive has been the concept, first introduced by Darwin, of evolutionary arms races. In his first book, written in 1912, Julian Huxley gave a graphic description couched in terms of the then-state-of-the-art naval military technology. "The leaden plum-puddings were not unfairly matched against the wooden walls of Nelson's day." Now however, "though our guns can hurl a third of a ton of sharp-nosed steel with dynamite entrails for a dozen miles, yet they are confronted with twelve-inch armor of backed and hardened steel, water-tight compartments, and targets moving thirty miles an hour. Each advance in attack has brought forth, as if by magic, a corresponding advance in defence." Likewise in nature, "if one species happens to vary in the direction of greater independence, the inter-related equilibrium is upset, and cannot be restored until a number of competing species have either given way to the increased pressure and become extinct, or else have answered pressure with pressure, and kept the first species in its place by themselves too discovering means of adding to their independence." Eventually, "it comes to pass that the continuous change which is passing through the organic world appears as a succession of phases of equilibrium, each one on a higher average plane of independence than the one before, and each inevitably calling up and giving place to one still higher."[8]

Huxley acknowledged the influence of Bergson on his writing, but one can see that, despite the Darwinian mechanisms, Spencer was at least as great an influence. No conflict here, for Bergson himself was influenced by Spencer. The crucial point is this: for all of Huxley's appeal to natural selection, there still seems to be a gap between the limited individual arms races leading to the perfection of certain adaptations (like thick shells for defense or sharp tools for penetrating thick shells) and those more sweeping changes necessary to produce human intelligence and other supposedly superior adaptations.

Why is this? Darwin was right: intelligence is indeed an adaptation. But what he failed to say is that intelligence is a *costly* adaptation. It demands brains, which in turn use up lots of energy. Really powerful brains like ours require large, regular intakes of highly concentrated protein: meat. Obtaining meat requires energy and skill. A cowlike existence on an abundance of low-energy fodder would not do. Hence, given the relativity of selection—which rewards simple survival and reproduction in the face of competitors, not the achievement of perfection—one has no guarantee that an arms race would take evolution in the direction of intelligence. Simple camouflage, for instance, might be worth a lot more than some sophisticated enemy-detecting apparatus.

The leap of faith required to produce human intelligence can be seen clearly in today's most ardent arms-race-based progressionist, Richard Dawkins. Though he would repudiate the term "leap of faith," Dawkins thinks that absolute biological progress is the inevitable result of evolutionary arms races, even though these seem only to produce a kind of relative perfection. He points out that, more and more, today's military arms races rely on computer technology rather than brute power, and—in the animal world—Dawkins translates this into bigger and bigger brains. In making his case, he refers to the notion of an EQ, encephalization quotient—a kind of cross-species measure of IQ that factors out the amount of brain power needed simply to get an animal to function (whales require much bigger brains than shrews because they need more computing power to control their bigger bodies) and then scales animals according to the surplus left over. Dawkins writes: "The fact that humans have an EQ of 7 and hippos an EQ of 0.3 may not literally mean that humans are 23 times as clever as hippos! But the EQ as measured is probably telling us *something* about how much 'computing power' an animal probably has in its head, over and above the irreducible amount of computing power needed for the routine running of its large or small body."[9]

Strengthening the bridge between the relative perfection produced by arms races and the absolute perfection that signifies real progress, Dawkins adds the notion of the "evolution of evolvability." Some new adaptations really do signify an absolute advance up the scale of perfection, spelling true progress, however defined. Against the thinking of his biggest rival in the realm of evolutionary popularization, Dawkins writes:

> Notwithstanding Gould's just scepticism over the tendency to label each era by its newest arrivals, there really is a good possibility that major innovations in embryological technique open up new vistas of evolutionary possibility and that these constitute genuinely progressive improvements. The origin of the chromosome, of the bounded cell, of organized meiosis, diploidy and sex, of the eucaryotic cell, of multicellularity, of gastrulation, of molluscan torsion, of segmentation—each of these may have constituted a watershed event in the history of life. Not just in the normal Darwinian sense of boosting evolution itself in ways that seem entitled to the label progressive. It may well be that after, say, the invention of multicellularity, or the invention of segmentation, evolution was never the same again. In this sense there may be a one-way ratchet of progressive innovation in evolution.[10]

Life's Solution

Although as a committed Christian, he abhors Dawkins's nihilistic vision of life and its meanings, the English paleontologist Simon Conway Morris—one of today's leading Darwinian evolutionary biologists—would agree with the sentiments of the last paragraph.

In one of his most popular books, *Wonderful Life*, Stephen Jay Gould singled out for special praise Conway Morris's work on the Burgess Shale, where soft-bodied, Cambrian organisms are well preserved in a shale deposit in the Canadian Rockies. At the same time, however, Gould used Conway Morris's findings as support for his own thesis about the nondirectedness of life's history. Gould argued that most of the Burgess Shale organisms went extinct, and pure chance determined which ones lost and which won. If we could turn the evolutionary clock back to the Cambrian (just over 500 million years ago), there would be no guarantee that subsequent history would turn out as it did. In Gould's vivid metaphor, "the tape of life" replayed would always be very different from the time before.

Conway Morris objects strongly to this conclusion, and recently has provided an extended, science-based argument intended to support his religious conviction that the arrival of humans on this planet was both highly improbable (in the sense that any life appearing at all was unlikely) and highly probable (in the sense that, once life commenced, intelligent beings were just a matter of time). Conway Morris's improbability-of-life arguments are fairly standard and, although updated by modern science, belong to a tradition going back at least to Whewell. More creative and interesting is the positive side of Conway Morris's argument, about the nigh-inevitable appearance of humanlike organisms once life commenced on earth.

Conway Morris's starting position is that only certain morphological shapes are capable of supporting functional life. He draws attention to the oft-noted absence of wheels in the living world. Given that wheels are such an efficient way of transporting loads, it seems very strange that, far from being ubiquitous, they are absent. Animals have legs, wings, and fins, and some slither, but not a one has wheels. The reason why wheels do not normally exist is very simple. Wheels need flat, hard surfaces to function properly, and such surfaces are rare in the natural world. "In the natural world as often as not, and especially on sea floors, this means acres of mud and other soft, sticky, substrates, ideal for getting bogged down."[11] Wheels are simply not adaptive.

Building on this point, Conway Morris adds the assumption that selection is forever pressing organisms to look for the few forms that can function in a given ecological niche. Wherever such a niche exists, sooner or later it will be occupied by the appropriate form—probably sooner rather than later, and probably many times. This directly challenges Gould's claim that, if you run life's tape more than once, you would get completely different results. Conway Morris draws attention to the way in which life's history shows an incredible number of instances of convergence—instances where the same adaptive morphological form has evolved again and again. A most dramatic example of convergence is that of saber-toothed tigerlike organisms: in North America placental mammals (real cats) appeared, whereas in South American marsupials (thylacosmilids) evolved. But in both places there clearly existed a niche for predators with catlike abilities and shearing/stabbing weapons, and natural selection found more than one way to occupy it. Indeed, it has been suggested that long before mammals and marsupials appeared, the dinosaurs might also have found this niche.

Conway Morris's repeated point is that this sort of thing happens over and over again, showing that the historical course of nature is not random but strongly selection-constrained along certain pathways and to certain destinations. From this, Conway Morris concludes that movement up the order of nature—the chain of being—is bound to happen, and eventually some kind of intelligent being (a "humanoid") inevitably emerges. We know from our own existence that a kind of cultural adaptive niche exists—a niche where intelligence and social abilities are the defining adaptations. Moreover, we know that this niche is one to which other animals have aspired (with greater or lesser success, as the many extinct species represented by Lucy's kin attest). We know of the kinds of features (eyes, ears, and other sensory mechanisms) that have been used by animals to enter new niches. We know that brain size has increased as selection presses organisms to ever new and empty niches. And finally we know that with this improved hardware have come better patterns of behavior—that is, more sophisticated software.

> If brains can get big independently and provide a neural machine capable of handling a highly complex environment, then perhaps there are other parallels, other convergences that drive some groups towards complexity. Could the story of sensory perception be one clue that, given time, evolution will inevitably lead not only to the emergence of such properties as intelligence, but also to other complexities, such as, say, agriculture and culture, that we tend to regard as the prerogative of the human? We may be unique, but paradoxically those properties that define our uniqueness can still be inherent in the evolutionary process. In other words, if we humans had not evolved then something more-or-less identical would have emerged sooner or later.[12]

The Bigger Picture

Conway Morris offers a view of life produced by a Christian, acceptable to a Christian. But he is a scientist, not a theologian; and precisely because he writes with the authority of a scientist, he has tended not (at least not yet) to put his thinking in a broader context. He is explicit that "a universe without an eschatological dimension is a universe that is incomplete." Moreover, his perspective is unambiguously postmillennial, in the sense that humans must work toward a brighter future. With respect to conservation, "We should question whether the biosphere is some malleable object for human exploitation, whether it be by pillage or modification." And we have obligations to the rest of nature, and should "take the claims of other sentiences, notably among the vertebrates and cephalopods, much more seriously." But he offers no theological or philosophical underpinning for these views.[13]

Others, thinking in parallel, have aimed for that bigger picture. Typical is the Catholic theologian John Haught, who speaks warmly of the vision and influence of Teilhard de Chardin, while trying to substitute the theology and metaphysics of the Anglo-American philosopher Alfred North Whitehead for the more Bergsonian vitalistic thinking of the French Jesuit. For Haught, following Whitehead, this world of ours is incomplete. Material being is the creation of God, and yet in some sense it is now separate from him, with its own status and nature. Because God has deliberately relinquished control, he cannot simply order or determine the outcome, though he can influence it and try to direct it. The world is not a machine that, once wound up, just spins around automatically in a deterministic manner, but rather seems almost to have life or potentiality of its own.

Echoing Richard Owen with his archetypes, Haught argues that "informational patterning . . . is a *metaphysical* necessity; for in order for anything to be actual at all, it must have at least some degree of form, order, or pattern." Haught denies that blind nature itself can produce this form: "It is unthinkable that novel events could arise only out of the fixed past." However, apparently God can push or guide creation in this way, when it is ready for such a move: "Although neo-Darwinian biologists often highlight what they take to be the aimlessness of evolution, if we step back and survey the life-process within its larger cosmic context, it is hard even for the most entrenched pessimist to discount altogether the obvious 'directionality' visible in the overall movement of the cosmos from simplicity to complexity. And to those of us who have been encouraged by faith to look for signs of promise in all things, it would seem egregiously arbitrary not to remark how, at any past moment in its history, the cosmos has remained open to surprisingly beautiful future outcomes."[14]

Haught writes movingly of the pain and suffering in this world, but he sees it as in some sense a necessity, as creation moves upward. For him, the focus is on the future. "Along with a general trend toward increasing complexity-consciousness, one may also observe, with the famous Jesuit paleontologist, that there has been an increasing 'centration' at the heart of the cosmic and evolutionary process. From the beginning, 'matter' has had a tendency to congeal around centers that function as attractors toward more complex and emergent levels of being. A progressively richer nucleation occurs in the atom, the eukaryotic cell, the central nervous systems of vertebrates, and most obviously human subjectivity." Haught elaborates:

New possibilities can arise only out of the region of time that we refer to as the future. And since the future is such a boundless reservoir of novelty, we cannot assume that, simply because it is not fully present to us now, it is reducible to bare nonbeing (as both philosophy and science have often implied). Because of its faithful and inexhaustible resourcefulness, we must concede to the future some modality of being. Indeed, the biblical version of reality's promise implies even that the future is the most real (although obviously not yet presently actualized) of all the

dimensions of time. The future claims the status of being emi-
nently real not only because it always shows up even after every
present moment has slipped away into the past, but ultimately
because it is the realm from which God comes to renew the
world.[15]

Haught speaks comfortably of his eschatological concerns, referring (as in the passage just given) to his biblical base. As he says elsewhere: "Biblical faith looks less towards a God transparently revealed
in present natural harmony and more toward a future coming of God
in the eschatological perfection of creation. It is especially this hopeful tone, and not just its sacramentalism, that can ground an ecological spirituality." Obviously, Haught does not use the literal language
of the apocalyptic books of the testaments, and indeed turns this
point to his advantage: "Like most theologians I believe that apocalyptic or millennialist forms of eschatology have to be thoroughly
demythologized in order to be taken seriously today. To take them literally is not to take them seriously." Yet, however Haught would char-
acterize himself technically, clearly his concerns are those of the
postmillennialist.[16]

This fits well with his respect for the similar-thinking Teilhard. "It
is necessary that all men of the earth should first reach the terminal
point of their evolutionary growth. This is a physical prerequisite.
Then it will be discovered that the perfection which lies beyond the
human—the ultra-human achievement—which neo-Humanism fore-
sees in evolution, is identical with the summit-point which all Christians expect under the term 'incarnation.'" Haught carries his own
commitment through at a practical level, fretting about ecological issues and problems and demanding that our present behavior be directed to a better future in these respects. "You've rightly characterized me as an eschatologist. I'm one of the growing number of
Christian theologians who, especially since the 1960's, after we got

over Karl Barth and neo-Thomism, believe that theology must be essentially eschatology, and that faith is inseparable from hope for the future—not just an otherworldly future but also the future of this world. Hence my interest in ecology."[17]

Being nice to nature is not just one option among many, something for the well-heeled middle classes to do in their spare time. The "Spirit of Life (the Holy Spirit) depends upon our feeling the power, complexity, and integrity of the biotic levels that support our own existence here and now. We might also argue, from an eschatological perspective, that if we lose touch now with nature's beauty we also risk losing our sense of the 'power of the future,' the renewing energy that faith perceives most explicitly in the resurrection of Christ." Or as Haught put it recently in a lecture commemorating the great physicist and natural theologian Robert Boyle, we must open ourselves to the "promise of nature."[18]

Moving beyond neo-Paley obsessions with design, we must focus on "nature's openness to the future." "In my view, the fact that the universe possesses a narrative character made possible by its inherent openness to the future is the greatest of wonders. It is a wonder that we generally take for granted, but it runs much deeper into nature than does design. The theme of 'nature as promise' harmonizes nicely with the eschatological orientation of biblical religion." This is a postmillennial eschatological orientation, because a biblically based faith "acknowledges the imperfection of the present state of creation and seeks to reshape the world, including the natural world, so that it will come into conformity with what it takes to be God's vision of the future."[19]

Self-Organization

Haught believes in a division of labor. Being a theologian, he leaves the science to others. But many people are disinclined to stay within such strict boundaries. Many of today's theistic evolutionists do not share the optimism of Conway Morris about the creative power of unaided Darwinism. Agreeing more with Gould, they do not see that a selection-driven process would necessarily reach upward to human life and beyond. In part, this skepticism arises because selection and its consequences, especially extinction, seem more like the dark side to creation rather than a bright light and hope for the future. Most people prefer a softer, more comfortable vision of the evolutionary process. Almost no one wants to belittle the problem of evil—especially the evil brought on by humankind—but true believers are reluctant to lay too much pain and suffering at the feet of God. A more integrative approach is needed, one which sees nature as a harmonious web rather than a parched battlefield of struggle and strife.

In the words of Keith Ward, Oxford Regius Professor of Religion: "On the newer, more holistic, picture, suffering and death are inevitable parts of a development that involves improvement through conflict and generation of the new. But suffering and death are not the predominating features of nature. They are rather necessary consequences or conditions of a process of emergent harmonisation which inevitably discards the old as it moves on to the new."[20] For many of today's theists, the problem with Darwinism is, in part, that the relativity of selection hardly guarantees the emergence of creatures made in God's image. Given the gaps in the arms-race argument, such doubters wonder just how selection could promote the evolution of evolvability. Likewise, they feel that Conway Morris is too optimistic about selection's ability to ferret out and fill the supposed empty niches waiting out there in morphological space. And we have not mentioned the assumptions that Conway Morris makes about there being ever-higher niches that take organisms up a progressive chain of being.

Thus, dissatisfied with unaided Darwinism, today's Christian evolutionists often turn to supplements. The physicist Robert J. Russell, in an updated version of Asa Gray's thinking, suggests that perhaps God puts in guidance at the quantum level, inaccessible to human eyes and concealed by the supposed randomness of phenomena at the lowest level we can discern, but there nevertheless. Through this, God is able to guarantee the upward processes of life.[21] Other theists find this solution worse than the problem it addresses. It introduces God into nature in a way that modern science excludes, and it is at best ad hoc and unsupported. Russell's suggestion is an example of what is known as a "god of the gaps" solution, which plugs in the Creator as a substitute for ignorance—on a par with Bergsonian vital forces. These theists insist that we must stay with a purely scientific solution. With this in mind, they are particularly attracted to the thinking of biological formalists, in the last century going back to the Scottish morphologist D'Arcy Wentworth Thompson and before him going back at least to the *Naturphilosophen* in the century before. This kind of thinking, recently revived and championed by those enthused by the modeling powers of computers, sees nature itself as containing creative and adaptive organization, which unfurls and manifests itself in the real world simply by executing the laws of physics and chemistry.

This so-called self-organization of brute matter has been referred to cleverly as "order for free." A popular example is the widespread phenomenon known as phyllotaxis—the spiral patterns that one sees ordering the parts (seeds and flowers) of many plants, for instance, the head of the sunflower.[22] One might think (Darwinians have thought) that this phenomenon is produced by natural selection, to maximize exposure to the sun while minimizing the amount of space needed to gain this benefit. In fact, one can show that such spirals have little or nothing to do with biological processes and are simply the results of the way in which seeds and other parts are produced. Seeds, such as those of the sunflower, are made in the center and then pushed outward. Simple formulas drawn from the branch of mathematics known as lattice theory show how these spiral effects are produced.

Inspired by such examples as these, critics of pure Darwinism argue that much of what we see in the organic world, especially much that we see as creative or organized in the organic world, is likewise a function of mathematical or physicochemical processes. For instance, the Christian evolutionists Bruce Weber (a chemist) and David Depew (a philosopher) refer to something they call "complex systems dynamics." This "includes the energetic driving force of far-from-equilibrium thermodynamics (that is, far from the equilibrium state in which there is no change in the energy of the system as a whole). This gives rise to matter-energy gradients and to internal structures, breaking symmetries and producing internal order and organization at the expense of increases in entropy (disorder and/or degraded energy) in the environment." Apparently, a "consequence of such a state of affairs is the appearance of nonlinear (nonadditive, potentially amplifying) interactions between components and processes in such systems through which organized structures emerge."[23]

Weber and Depew go on to argue that these processes can produce life from nonlife, and complex life (up to our own species) from simple life. They stress that these processes are not to be construed as rivals to natural selection, and are certainly not in opposition to it. Rather, just as Mendelian genetics complemented and fleshed out the processes of selection as understood in the 1930s, so also self-organization, or what they prefer to call system organization, further fleshes out the processes of selection as we understand them today: "Although selection and self-organization can be conceptualized as interacting in a variety of logically possible ways, these two principles can and should be thought of as complementary *phenomena.*"[24]

An Ecological Vision

Holmes Rolston III, a Presbyterian minister, professional theologian, recent winner of the Templeton Prize for advances in religion, and, in the view of many, today's leading philosopher of the environment, has tried to tie these ideas all together. A totally committed evolutionist, Rolston repudiates the philosophical meaninglessness of Darwinism and rejects the scientific adequacy of natural selection. So what do we put in their place? Rolston believes that we need from nature a guarantee of order and upward direction. In the age of DNA, vital forces will not do; they are entirely against the spirit of modern science. An alternative is needed, and (in line with self-organization thinking) Rolston finds the solution to life in a kind of physical ratchet effect: molecules organize themselves spontaneously into ever-more complex configurations. These then get incorporated into living beings, bringing on a general move upward, with apparently no return possible. "But to have life assemble this way, there must be a sort of push-up, lock-up effect by which inorganic energy input, radiated over matter, can spontaneously happen to synthesize negentropic amino acid subunits, complex but partial protoprotein sequences, which would be degraded by entropy, except that by spiraling and folding they make themselves relatively resistant to degradation. They are metastable, locked uphill by a ratchet effect, so to speak, with such folded chains much upgraded over surrounding environmental entropic levels. Once elevated there, they enjoy a thermodynamic niche that conserves them, at least inside a felicitous microspherical environment."[25]

From all of this, we get a kind of biological progress. Admittedly, every "form of life does not trend upslope." But overall, the movement is upward, culminating in humankind, God's special creation. This upward struggle of evolution parallels the upward struggle of our species. "What theologians once termed an established order of creation is rather a natural order that dynamically creates, an order for creating. The older and newer accounts both concur that living creatures now exist where once they did not. But the manner of their coming into being has to be reassessed. The notion of a Newtonian Architect who from the outside designs his machines, borrowed by Paley for his

Watchmaker God, has to be replaced (at least in biology, if not also in physics) by a continuous creation, a developmental struggle in self-education, where the creatures through 'experience' become increasingly 'expert' at life." Not that this is cause for worry or alarm. "This increased autonomy, though it might first be thought uncaring, is not wholly unlike that self-finding that parents allow their children. It is a richer organic model of creation just because it is not architectural-mechanical. It accounts for the 'hit and miss' aspects of evolution. Like a psychotherapist, God sets the context for self-actualizing. God allows persons to be imperfect in their struggle toward fuller lives . . . and there seems to be a biological analogue of this. It is a part of, not a flaw in, the creative process."[26]

Rolston argues strenuously that to lead a full life, we must pay attention to the world around us—animals and plants—and try meaningfully and consistently to preserve and cherish the biosphere. In this, he is at one with the secular postmillennialist Edward O. Wilson, believing that values—their character and their foundation—arise from the evolutionary process. Nature itself is neither moral nor immoral. But "amoral nature is fundamentally and radically the ground, the root out of which arise all of the particular values manifest in organisms and ecosystems. This includes all human values, even though, when they come, human values rise higher than their precedents in spontaneous nature." And as with Wilson, the living world and its care is the highest value of them all. "Environmental ethics . . . is the most altruistic, global, generous, comprehensive ethic of all, demanding the most expansive capacity to see others, and this now especially distinguishes humans." Roylston stresses that this is not a reductionistic naturalized ethics. Rather, "it is naturalized ethics in the comprehensive sense, humans acting out of moral conviction for the benefit of nonhuman others. There is a widening sense of shared values, including values produced in the evolutionary genesis."[27]

Like Haught, Rolston is loath to tie his thinking to any literal reading of apocalyptic visions, but his theology directs him to think about the future.

I do pay attention to the future. If a is the cause of b, sometimes we look to the past and understand b in terms of a. Tides are caused by the gravity of the moon. But in historical and narrative explanations we just as often understand a in terms of b. Plymouth Rock is not so much understood by looking backward as by looking forward. In terms of natural history, trilobites are the outcome of quarks, but what is going on is not so much illuminated by understanding quarks as by seeing the evolutionary story and what new possibilities are opening up with trilobites, which result in a lot more of interest to follow. Stories have beginnings, middles, and ends—remember. Nature is to be understood as much by what lies ahead as by what lies past. Here I do pay attention to the so-called theologies of hope. We are not so much interested in where we have come from as in where we are going, or what the choices are that we have about what kind of future we ought to have. Past is fact; future invites ethical obligation, often based on religious vision.[28]

For Rolston, nature inclines toward progress, humans are part of this process, and we have an obligation—as Christians—to look to and work for a brighter future. "The abundant life that Jesus exemplifies and offers to his disciples is that of a sacrificial suffering through to something higher." For this reason, "the aura of the cross is cast backward across the whole global story, and forever outlines the future."[29] In the terms of today's liberal theology, we have a postmillennial eschatology.

12

Earth's Last Days?

He approached the bed, knowing what he would find. The indented pillow, the wrinkled covers. He could smell her, though he knew the bed would be cold. He carefully peeled back the blankets and sheet to reveal her locket, which carried a picture of him. Her flannel nightgown, the one he always kidded her about and which she wore only when he was not home, evidenced her now departed form.

His throat tight, his eyes full, he noticed her wedding ring near the pillow, where she always supported her cheek with her hand. It was too much to bear, and he broke down . . . With wooden fingers he removed his clothes and let them fall to the floor. He climbed into the bed and lay facedown, gathering Irene's nightgown in his arms so he could smell her and imagine her close to him.

And Rayford cried himself to sleep.

*Y*ou might think that the missing wife of Rayford Steele (captain, Pan-Continental Airlines) has run off with the milkman. But you would be wrong. The time has come for God to set things in motion. Already, Israel and the Jews have been saved miraculously from the hordes without. Now comes the next stage. The absent Irene Steele has—together with other true believers—been raptured up to heaven. She has cashed in her ticket to everlasting salvation. Her grief-stricken husband—who failed to take seriously the warnings of the genuine Christians and who, at the moment of the ascent, was lusting after a comely flight attendant—is one of the poor wretches left behind to gird up their loins and fight the Antichrist.

Premillennial dispensationalism is alive and well in the United States of America today. The sales figures of *Left Behind: A Novel of the Earth's Last Days,* by prophecy scholar Dr. Tim LaHaye and professional writer Jerry B. Jenkins, are of a magnitude to make commercial publishers swoon and secular authors turn an unsightly shade of green. The full series of novels has sales over 60 million copies, and counting. And that's just the start. Apart from webpages and newsletters ("updates on the newest releases, inspirational articles, trivia quizzes, chances to win great *Left Behind* prizes") and Bible covers, T-shirts, mugs, magnets, journals, and key chains, we have a *Left Behind* series for kids ("As the world falls in around them, Judd, Vicki, Lionel, and the others must band together to find faith and fight the evil forces that threaten their lives!"), with 10 million copies sold, a *Left Behind* military series, a *Left Behind* political series, and, for those of us differentially gifted, a series of graphic novels, which "maintain the style of a comic book, but have the sturdiness of a novel."

You might object that *Left Behind* is all fiction and nothing to get so excited about. Many people enjoy the books and movies about James Bond, without believing that real-life agents are like 007; even fewer think they themselves will ever own the kind of futuristic automobiles that Bond drives or possess the beautiful women who cluster around him like flies around a jam pot. The analogy has some merit, but make no mistake: the writers of the *Left Behind* series are deadly serious, and so are many of their readers. The books and the spin-off kitsch represent a whole industry devoted to making devotees realize that this is fiction with a purpose—fiction that is telling us how it really is, or rather how it really is going to be. Books, videos, and seminars conducted throughout the nation reinforce the message that the premillennial picture painted in *Left Behind* is no chimera—no silly fantasy—but a real account of the fate that awaits believers and unbelievers alike.

Flood Geology

To understand this latest upsurge of premillennialist fervor, let's turn
back for a moment to the early part of the century, when three views
of the history of life were available to Christians. In the Scopes Trial,
William Jennings Bryan favored an "ages" interpretation of Genesis,
where the six days of creation are interpreted as six very long periods of time. Others (probably the majority, given that the view was
adopted in the Scofield Bible) endorsed the "gap" interpretation,
where a long unrecorded period is inserted into the week, generally after the physical creation and before life as we know it appears. Really
strict literalists continued to insist on a week's span of twenty-four-
hour days, the final one of which was the Sabbath on which the Lord
rested. For them, everything happened about 6,000 years ago. At
the time of the Scopes Trial, this "young-earth creationism" was a minority position, associated primarily with Seventh-day Adventists.
Founding prophetess Ellen G. White had visions that God did do his
work in six literal days and that the seventh (a Saturday) should therefore be kept sacred by us. One had no wiggle room for any modifications in the most strict interpretation of Genesis.

In the first half of the twentieth century, the most ardent promoter
of the literal-day interpretation was a Canadian-born Adventist,
George McCready Price.[1] Constructing a theory of "flood geology,"
the long-lived Price produced one book after another supporting
White's visions. Since God undoubtedly did his creative work efficiently and quickly, Price's basic problem was to explain the disturbed and fragmentary nature of the fossil and geological record as
we find it today. Why are there traces of great upheavals and destruc-
tions and of general earthworks like the Himalayas and so forth?
Neatly fusing a somewhat idiosyncratic view of earth's history with
his reading of Genesis, Price (following White) blamed it all on the

universal flood through which Noah and his family alone survived. The flood messed up God's perfect creation. The seemingly progressive fossil record is an artifact of this flood. Those organisms highest in the record were not the most recent ones to evolve but those best able to climb the highest mountains—at which point they too were submerged and drowned. "The larger animals and man would flee to the hill-tops from the rising waters, and, when finally engulfed, would be simply drowned, and not immediately covered with earthy deposits."[2]

For Price, this worldview was all bound up with the premillennialism of the Adventists. The flood at the beginning corresponds to the forthcoming Great Tribulation at the end. "Surely it is useless to expect people to believe in the predictions given in the last chapters of the Bible, if they do not believe in the record of the events described in its first chapters." Price was not that bothered by evolution as such. Geology and the flood were what really counted to explain the diversity of species we see today. He was prepared to allow a fair amount of organic change after the animals came out of the Ark, since clearly Noah did not have room to take two specimens of every living kind. He was inclined, however, to think that such change was the direct work of the devil rather than any lawbound natural process like selection.[3]

The racial differences among humans are recent examples of evolution, according to Price, and they reflect God's curse on some of Noah's sons for their sinful weakness.

> The poor little fellow who went to the south
>   Got lost in the forests dank;
> His skin grew black, as the fierce sun beat
> And scorched his hair with its tropic heat
>   And his mind became a blank.[4]

Price's geological literalism was part and parcel of a moral attitude toward the world and its human denizens. "Do you know that the theory of evolution absolutely does away with God and with His Son Jesus Christ, and with His revealed Word, the Bible, and is largely responsible for the class struggle now endangering the world?"[5]

Somewhat surprisingly, young-earth creationism became the standard position in the second half of the twentieth century.[6] The background to this development was an intensification of the culture war between the two sides of the reason/faith, postmillennial/premillennial dichotomy. Although in the years directly after the Scopes Trial evolutionary theory finally came to professional maturity, in other respects evolutionary thinking was badly served during that time. By around 1930 the intense public attacks of the fundamentalists had faded away—there had been too much laughter and scorn in Tennessee for them to keep their confident profile. Yet the damage had been done, especially in the realm of education in America. Textbook publishers aim to sell large quantities of their products in all states of the Union, and hence they tend to go for the lowest common denominator. Fairly quickly, biology texts were gutted of what little evolution they contained, and even if Genesis was not offered as an alternative, students were kept ignorant of what was happening in the scientific world at large.

This changed dramatically at the end of the 1950s, when a thoroughly progressionist version of neo-Darwinism became common currency, both in the public arena and increasingly in schools. In part, this change occurred because the now-confident evolutionary biologists were themselves making their beliefs very public. Celebrations in 1959 of the centenary of the *Origin* were vehicles for much prominent advertising of modern thinking on the histories of organisms. The highlight of a week-long jamboree at the University of Chicago—a week to which many schoolteachers were invited—was a sermon in the campus chapel by Julian Huxley, where he preached the virtues of a progressionist evolutionary humanism.[7] In part, changes favorable to evolutionary thinking—and to all scientific endeavors—came about in reaction to the Russians' success in putting the satellite *Sputnik* into orbit. America perceived itself as failing in the cold war and its associated arms race, and the nation's leaders were determined to upgrade the quality of American science, including science education. New textbooks, subsidized by the government, included full and approving discussion of the updated state of Darwinian evolutionary theory. The existence of these texts endorsed the virtues and necessity of progress. Even the text titles echoed this philosophy: *Biological Science: Molecules to Man.*

Here was one side of the American equation. But the post–World War II period in America also turned out to be a boom period for premillennialists and their followers. The war itself had convinced many that happenings of great significance were afoot, and the exploding atom bombs in the final days of the conflict were uncomfortably close to the burning fires and forces of destruction that Daniel and John had envisioned. Then, immediately following the physical battles came the cold war, not to mention another very real war in Korea. Americans felt threatened by forces of evil on the other side of the Iron Curtain—not that America itself was free of sin, given such abominations as girls in pants, drive-in movies, and rock and roll.[8]

The founding of the State of Israel in 1948 was clearly part of God's forecast during the last days, and even the formation of the Common Market was a portent of world domination by the Antichrist. Modern technology and mass communication were especially feared, because of the opportunities they opened up for the Antichrist and his minions to make their bid for world dominance. (Later, computers would become objects of suspicion.) Paradoxically, exploiting these same media, doomsday prophesiers flourished, including some of great influence and prominence. In the 1950s and 1960s the most important evangelist of the century, Billy Graham, frequently couched his sermons in a premillennial context: "The whole world is hurtling toward a war greater than anything the world has ever known before."[9] Even the obtuse must see that the new weapons of mass destruction are precisely those referred to in 2 Peter 3:10. "But the day of the Lord will come as a thief in the night; in which the heavens shall pass away with a great noise, and the elements shall melt with fervent heat, the earth also and the works that are therein shall be burned up."

In the context of religious and moral sentiments such as these, two men took up Price's flood geology and translated it into terms that made it the touchstone for creationist thinking (the name by which, in the second half of the twentieth century, fundamentalist thinking about origins became known). *The Genesis Flood: The Biblical Record and Its Scientific Implications* appeared in 1961, authored by John C. Whitcomb, a Princeton-educated theologian, and Henry M. Morris, a hydraulic engineer. All of the old arguments and many new ones were trotted out, to show that the Bible, taken literally, is a good guide to the history of the earth and the life one finds there. Our globe is between 6,000 and 10,000 years old; the creation was done rapidly in a matter of days; life appeared miraculously; and sometime thereafter a worldwide flood wiped out almost everything on the earth, save only a handful of people in the Ark.

According to the creationists, the fossil record, properly understood as reflecting the sequential way in which the flood killed off living beings, is the biggest piece of evidence supporting this view. Conversely, several instances of fossils have been found in places that quite negate the idea of evolution. A human skull was found down in the coal deposits, for example. Dinosaur and human footprints were found that shared the same stratum. The human footprints were often of very great size, reminding one of the claim in Genesis 6:4 that there were "giants in those days."[10]

Again and again Morris (who became the leading spokesman for this new creationism) reaffirmed his commitment to a premillennialist reading of holy scripture. Like Price, his was no coincidental commitment but the *raison d'être* for his obsession with origins. We are living in a time before the Second Coming of Christ, who will then rule for a thousand years. "The millennium—the thousand-year reign of Christ with His resurrected saints, as outlined in Revelation 20:1–7—can only be future if it is literal." Hence, "the events described in the Book of Revelation are real events, just as sure to take place in future history as the events in Genesis, and all the other historical books of Scripture, took place in past history."[11]

In *Genesis Flood*, the significance of the deluge for this reading of future events was stressed strongly. Whitcomb and Morris, like Billy Graham, referred to the passage in II Peter where the writer warns that God's day "will come as a thief in the night" and when "the elements shall melt with fervent heat, the earth also and the works that are therein shall be burned up." They made much of the fact that this warning comes immediately after reference to the devastating effects of the God-caused flood—"Whereby the world that then was, being overflowed with water, perished"—and they commented that here "the Flood is used as a type and warning of the great coming worldwide destruction and judgment when the 'day of man' is over and the 'day of the Lord' comes."[12] Peter noted that no one would believe his warnings and he would be scoffed at; such indeed is the present state of things, Whitcomb and Morris pointed out.

As always with premillennialism—with postmillennialism too, for that matter—physical events, past and future, are only part of the story. Their other concern is morality. What must we do to be saved? From the first, with dogged determination, Morris has warned of the moral issues at stake here. In a textbook that he coordinated, the preface stated flatly that evolutionism corrupts proper teaching. "Not only is this system inimical to orthodox Christianity and Judaism, but also, as many are convinced, to a healthy society and true science as well." In a more recent book Morris singled out a "notorious Darwinian philosopher Michael Ruse," apparently a well-known "atheistic humanist," as contributing to the moral rot. "It is rather obvious that the modern opposition to capital punishment for murder and the general tendency toward leniency in punishment for other serious crimes are directly related to the strong emphasis on evolutionary determinism that has characterized much of this century." Whereas the focus of postmillennialist moral exhortations is on societal or general improvement—save the rain forests and that sort of thing—the focus of the premillennial moral exhortations is on individual purity, to avoid punishment. Those who fail to achieve it will be damned, not socially engineered into reformation.[13]

Premillennialists stress again and again the need to spread the gospel and, Moody-like, to save as many souls as possible before the end comes. As the decades went by, this need grew ever greater: who could rest easy in a world where America's leaders were assassinated, flower children gathered for public love-ins, the body count in Vietnam rose ever higher, and youthful protesters were clubbed and shot on college campuses and city streets? Even the triumphs of technology were ambiguous. Was not the space race itself an "open defiance of God's plan"?[14] Remember Psalm 115: "The heaven, even the heavens, are the Lord's: but the earth hath he given to the children of men."

Following in the wake of the countercultural revolution, the 1970s especially was a time when premillennialism rode high. The decade's bestseller was Hal Lindsey's apocalyptic tract *The Late Great Planet Earth*, a work that covered all of the usual territory with something nigh to enthusiasm as it forecast the horrors to come. One

could easily dismiss such a crude literary production and sneer at
Lindsay's description of the Antichrist as "the Weirdo Beast" and of
the rapture as the "Ultimate Trip" (though even the hardened cynic
must admit to something ethnically touching about the suggestion
that when the waters of the world have turned to blood, humankind
will be able to survive for a while on Coca-Cola). Less easy to dismiss
was the commercial appeal of *Planet Earth*. Published in 1970, it had
sold 9 million copies by 1978 and 28 million by 1990. Subsequent
works from the same author also spent many weeks on the *New York
Times* bestseller list.

Joining in this premillennialist movement to spread the truth of the
apocalypse and save misguided and ignorant souls were the creationists. Henry Morris and others (including Tim LaHaye) banded together
in creationist organizations, writing and publishing books, lecturing,
starting a college, and doing their bit to combat the teaching of evolution in schools and to provide alternatives in line with the vision of
Genesis Flood. Legally, things had moved on from the days of Scopes.
The U.S. Supreme Court had ruled that the teaching of evolution
cannot be excluded from schools—and so now the battle turned to
introducing creationism as an alternative interpretation of life's history.
Because the U.S. Constitution separates church and state, the Genesis
account had to be sanitized and presented as justifiable on scientific
grounds alone. The Bible itself had to be kept out of the discussion, or
at least it had to be kept out of the proof. Later, it could be shown that,
as it just so happens, the conclusions of science and the Bible coincide.

Thus was born scientific creationism or creation science, a supposedly secular version of flood geology—same arguments, same conclusions, but less talk of giants and tribulations and so forth. A very
successful production along these lines was *Evolution? The Fossils Say No!*
by Duane T. Gish, a Berkeley Ph.D. and former biochemist at the
Upjohn Pharmaceutical Company. More than any other person in the

movement, he became skilled at debating with evolutionists, often before large college audiences. "Creationism's bulldog" frequently reduced his scientific opponents to speechless outrage through clever
linguistic tricks and agile application of pertinent facts. Thomas
Henry Huxley would have exploded with fury and left the podium—
and that would have suited Gish just fine.

Scientific Creationism

The debate quickly moved into the political arena. The then governor
of California, Ronald Reagan, was on record as saying: "Everything is
falling into place. It can't be long now. Ezekiel says that fire and brim-
stone will be rained on the enemies of God's people. That must mean
that they'll be destroyed by nuclear weapons."[15] Regular efforts were
made to introduce bills mandating the teaching of creation science
alongside evolution in the biology classes of state-supported schools.
In 1981 this effort finally met with success. After virtually no discussion, such a bill passed the two elected houses in Arkansas and was
sent up to the governor, Frank White. He was in office because—for
the first and last time in his life—the previous governor, Bill Clinton,
had failed to mend his political fences and was defeated in his run for
reelection. The new governor—who lost his job the next year when
Clinton regained office—was as unqualified for the post as he was
surprised at getting it, and he signed the bill without a thought. Some
fifty-six years after the clash in Tennessee, the scene was set for another court confrontation.

Q. What is your personal belief in the existence of a God?
A. I would say that today my position is somewhere between deist—
    that's to say in believing in some sort of, perhaps, unmoved
    mover—and agnosticism. In other words, don't really know.

I mean, I'm a bit like Charles Darwin in this respect. Some days I get up and say, "You know, I'm sure there must be a cause." And then other days I say, "Well, maybe there isn't after all."

Q. There must be a cause?

A. There must be something that—There must have been something originally.

Q. The term "cause," why do you use that in relation to your concept of God?

A. I'm talking about in the sense of some sort of ultimate religious sort of reason. It doesn't necessarily mean cause in the sense of physical cause. It could well be final cause or something like this.

Q. Is your conception of a God some sort of world force? Is that one way you would describe it?

A. As I say, I don't say my conception of a God is some sort of world force. My conception is, perhaps, sometimes there is more to life than what we see here and now.

Q. But you did tell me in your deposition that your conception of God would be that there might be some sort of, quote, world force?

A. There might be because, as I say, I'm not even an expert on my own beliefs in this respect.

Q. Do you have a personal belief as to whether a creator, in whatever form, had a hand, figuratively speaking, in creating the universe, life or man?

A. Not really. It's all so foggy to me.

Given that performance, I suspect that the atheistic humanists were no more keen to claim me as one of their own than were the creationists. This exchange occurred in testimony that I gave in Little Rock, Arkansas.[16] As a historian and philosopher of science who specializes in Darwinian issues, I was used by the American Civil Liberties Union as an expert witness to challenge the constitutionality of Arkansas's law requiring the "balanced treatment" of evolution and scientific creationism in publicly funded schools. My role was to testify on the nature of science and on the nature of nonscience, religion in particular. I was to show that evolution falls on one side of the line—the constitutionally acceptable side of the line—and that creationism, whatever it is called, falls on the other, constitutionally unacceptable side of the line.

Notwithstanding this somewhat bumbling exposition of my own religious views, the Arkansas law was indeed declared unconstitutional. In the judge's opinion, creation science is religion and not genuine science.[17] Relying on my testimony and that of others, he ruled that the "essential characteristics" of what makes something scientific are:

1. It is guided by natural law;
2. It has to be explanatory by reference to natural law;
3. It is testable against the empirical world;
4. Its conclusions are tentative, that is, are not necessarily the final word; and
5. It is falsifiable.

In his opinion, creation science fails on all counts. Evolution wins!

Well, not exactly. For starters, a number of philosophers—far more distinguished than I—took immediate umbrage at the suggestion that one can draw a sharp line between science and the rest of human activity. They argued that—for all that their conclusion would hardly have done the job in court—creation science is far from unfalsifiable. The system is open to check, and it is false. But apart from these criticisms—and like most of my breed, I happily responded to my tormentors, believing that one is far better off being criticized than ignored—what happened in Arkansas did little or nothing to curb Americans' enthusiasm for creationism, and (the converse) its distrust of evolution. Many people continued, and still continue, to buy into one form or another of creationism—and even if they were undecided or ambiguous, they agreed with Reagan that if evolution is taught, then the same courtesy should be extended to creationism. In 15 to 20 percent of American schools today, creationism is taught in biology classes.[18]

This should come as no surprise. In 1991, ten years after Arkansas, a Gallup poll found that 47 percent of Americans claimed to be creationists, and another 40 percent subscribed to some kind of theistic (that is, guided) evolution. These figures have remained very stable through the last decade and are confirmed by the surveys of other organizations. A 2001 Gallup poll reported that 45 percent thought that God created humans as they are now, 37 percent let some kind of guided evolution do the job, and 12 percent put us down to unguided natural forces (a percentage up 3 percent from 1982 but chiefly at the expense of the "don't knows," down from 9 to 6 percent). A 2001 National Science Foundation survey on science literacy found that 47 percent of Americans think that humans were created instantaneously (while 52 percent think that humans and dinosaurs coexisted).

Obviously, people like this tend to be at the more politically and socially conservative end of the spectrum. "Perhaps most notably, those for whom religion is an important part of life (those who attend religious services every week) are far more likely to prefer the theory of creationism than are those who attend church less often (80% versus 47%, respectively)."[19] Creationists tend to live in the South and the West, where Baptists and Methodists predominate. But as always when it comes to religion and politics, one should be wary of strong general claims. There are certainly those in the North who are not keen on evolution. And likewise, although premillennialism dominates among those who doubt evolution, one should not assume that everyone who rejects evolution is therefore an ardent premillennialist. This has never been the case. Bryan, for one, did not think Jesus was coming before the millennium. Still, according to some estimates there are at least thirty million hard-core premillennial dispensational-ists in America today, and many fellow travelers. Not a few have some trouble distinguishing the supposed fact of Revelation from the sup-posed fiction of the *Left Behind* series.[20]

Intelligent Design

The last decade of the twentieth century introduced us to the most recent set of Christian-based Darwin critics. The "intelligent design" (ID) movement started around 1990, when the Berkeley law professor Phillip E. Johnson published a little book, *Darwin on Trial.* Highly crit-ical of modern evolutionary thought, Johnson trotted out arguments that have been familiar since the time of the *Origin*—the inadequacy of natural selection, gaps in the fossil record, the problem of life's ori-gins, and so forth. Familiar though the arguments may have been, Johnson knew how to mount a prosecution, and it soon became clear that his book was a more polished work than the run-of-the-mill antievolutionist tract. By the time he had finished, Johnson felt justi-fied in finding the old evolutionist guilty, and led him away in chains.

Darwin on Trial sold like hot cakes. Naturally, it therefore attracted much negative attention as well. One justifiable criticism of Johnson was that he behaved too much like a lawyer—he concentrated on the offense and paid no heed to defense. As Thomas Kuhn had rightly pointed out in his *Structure of Scientific Revolutions,* scientists do not give up on old paradigms until there are new paradigms to substitute. What could Johnson offer as a substitute for Darwinism? An answer came later in the decade. In *Darwin's Black Box,* the Lehigh University biochemist Michael Behe invited his readers to focus on something that he called "irreducible complexity"—a system where all of the parts are intricately matched together in such a way that the system breaks down if any part is removed. "An irreducibly complex system cannot be produced directly (that is, by continuously improving the initial function, which continues to work by the same mechanism) by slight, successive modifications of a precursor system, because any precursor to an irreducibly complex system that is missing a part is by definition nonfunctional." Behe argued, surely with some truth, that any "irreducibly complex biological system, if there is such a thing, would be a powerful challenge to Darwinian evolution. Since natural selection can only choose systems that are already working, then if a biological system cannot be produced gradually it would have to arise as an integrated unit, in one fell swoop, for natural selection to have anything to act on."[21]

But are there such systems? Behe thought that there are. He highlighted an example familiar to those who work on microorganisms. To move about, some bacteria use a cilium, a kind of whip that contains its own power unit and works by a paddling or rowing motion. Others go a different route, using a flagellum (another whiplike appendage) as a kind of propeller. "The filament [the external part] of a bacterial flagellum, unlike a cilium, contains no motor protein; if it is broken off, the filament just floats stiffly in the water. Therefore, the motor that rotates the filament-propeller must be located somewhere else. Experiments have demonstrated that it is located at the base of the flagellum, where electron microscopy shows several ring structures occur." You need a connector between the external part of the flagellum (the filament) and motor, and lo and behold nature provides it: a "hook protein" is right there to do the job as needed. The system is highly complex in itself, and added to that is the fact that the motor has its own energy source. Energy is created continuously by the cell, rather than, more conventionally, by drawing on energy stored in various complex molecules.[22]

To Behe, none of this could have evolved or appeared in a gradual way, bit by bit. It all had to be there, up and running, in one creative act, and this required conscious design and flawless execution. Darwinism does not work that way, and so we must look for another explanation. Within the confines of naturalism, the only account on offer would be a "hopeful monster who luckily gets all of the proteins of the right nature in the right order at once" by pure chance. Such a silly explanation is no real answer. In Behe's view, irreducible complexity condemns naturalism and leaves design by an intelligent agent as the only plausible alternative.

In short succession, backing Behe's empirical argument, came conceptual arguments from the philosopher-mathematician William Dembski. Initially, his aim was to show us why we pick out and highlight things we call "designed." Dembski stressed that three notions underlie the concept of design: contingency, complexity, and specification. Design has to be contingent. Things that follow by blind law give no evidence of design. The ball falls and bounces. No design here. Second, design has to be complex. The number 2 followed by 3 does not catch your attention. But the succession of prime numbers up to 101 make you think that something interesting is going on. Finally, design demands a certain independence or specification—you cannot confer the criteria for design after the fact. Drawing a bull's-eye around an arrow that has already landed is not design. But an arrow that hits a bull's-eye that was specified in advance makes you think that the arrow's position is not random.[23]

Putting these criteria into action, Dembski introduced what he calls an "explanatory filter." I will illustrate it using my own examples, rather than the more mathematically oriented examples of Dembski. We have a phenomenon, and we want to know its cause. Did it happen necessarily as a consequence of the laws of nature, or is it contingent? Is it possible that it might not have happened? The planets circle the sun endlessly because of the blind laws of nature. No puzzle, no design.

But what about something that is contingent—that did not necessarily have to happen? Take a mutation, which may be predictable over large numbers and a huge expanse of time but occurs sporadically at the individual level. One has no reason to think that at this level such a mutation is necessary. But is it design? Not automatically. Consider a mutation to the hemophilia-causing gene, an affliction that apparently crippled the extended royal family of Europe. Such a mutated gene does not qualify as complex, for it leads to breakdown rather than otherwise. It results from chance or randomness, not design. The hemophilia gene is purely accidental.

What about something that is both contingent and complex? An example would be a particular pattern of mountains in a range, or a peculiar kind of rock formation that seems especially beautiful or weird. You obviously cannot deduce the patterns or formation directly from the laws of the physical sciences, even if you include geology with physics and chemistry. And the Rockies do not seem to represent a breakdown of order in the way that the hemophilia mutation among European royalty was. Should we therefore now speak in terms of design? Not yet. Nor should we speak of design for a pretty piece of rock, or an uncut gem. In the case of the Rockies, their patterns and arrangements were not specified in advance; they are the ad hoc results of forces working in nature, and not the result of the conscious intention of an intelligence. A rough diamond may be very valuable, but it is still just a natural object, with no prespecified design.

Mount Rushmore and the Hope Diamond are another matter. The presidents' profiles, the cut and polished stone, were mapped out before they were created. They are objects of design. Likewise, the microscopical biological apparatuses and processes discussed by Behe. In addition to being contingent and irreducibly complex, they show evidence of design, for they do exactly what is needed for the organism in which they are found. An intelligent agent studying the flagellum and its actions could have specified just what was needed by way of a motor. Consequently, this apparatus fits the third condition for design, which is specification. And so, having survived Dembski's explanatory filter, Behe's microbes are properly considered the product of real design.

Dembski was not writing in a theological vacuum. His arguments provide a neat solution to a major problem facing intelligent design theory: why, if a designer was involved in producing the very complex and wonderfully functioning organisms we see around us, did he (or it) not avoid producing the very simple yet devastating effects? Thanks to Dembski's analysis, the designer gets points for the flagellum motor (the product of design) and yet is off the hook for the hemophilia mutation (the product of chance). But Dembski's argument does not as such prove that there must be a designer. Or rather, his argument does not disprove the existence of special laws that might produce what he characterizes as design—the law of natural selection, for example.

Recently, Dembski has gone more on the offensive, invoking what are known as "no free lunch" mathematical theorems, arguing that "these theorems cast doubt on the power of the Darwinian mechanism to account for all of biological complexity." The basic thrust of this mathematical analysis is to say that you cannot get more out of a system than you put into it. Garbage in, garbage out. Dembski himself concedes that "the No Free Lunch Theorems are book-keeping results," to which he then adds: "But book-keeping can be very useful. It keeps us honest about the promissory notes our various enterprises—science being one of them—can and cannot make good. In the case of Darwinism we are no longer entitled to think that the Darwinian mechanism can offer biological complexity as a free lunch."[24]

Essentially, what Dembski says is that genuine design—phenomena that conform to his above-stated criteria—simply cannot be created by nondesign mechanisms. Design out implies design in. Bookkeeping admittedly, but book-keeping sufficiently powerful to make impossible a fully selection-created, living universe, in Dembski's view.

Naturalism

Creationists have welcomed the intelligent design movement as a sophisticated addendum to their antievolutionary view of life's history. Critics have likewise placed it in this tradition. I myself have referred to it as "creationism lite." Asked about how he thinks the crucial change occurs, Michael Behe responded: "In a puff of smoke!" Pressed with the question, "Do you mean that the Intelligent Designer suspends the laws of physics through working a miracle?" he replied, "Yes."[25] Others, Dembski particularly, prefer not to use the word miracle, because they think this word has religious connotations that are inappropriate in a scientific context. They deny that they are in a direct line back to Morris and earlier folk.

However it may be qualified, the point of the intelligent design movement is to promote the intellectual respectability of interventions outside the natural order of things. And we can certainly take with a pinch of salt the movement's claims about the absence of theological links and motives. Dembski has stated bluntly that "a scientist in trying to understand some aspect of the world, is in the first instance concerned with that aspect as it relates to Christ—and this is true regardless of whether the scientist acknowledges Christ." He added later: "Intelligent design is just the Logos theology of John's Gospel restated in the idiom of information theory." In the language of the Good Book: "In the beginning was the word, and the word was with God, and the word was God" (John I:I).[26]

Yet it would be a mistake simply to categorize the intelligent design movement as creationist without qualification. Many if not most of the leaders of the ID movement subscribe to, or at least are open to, some form of evolution. Behe: "I'm an 'evolutionist' in the sense that I do think natural selection explains some things . . . But from what I see, the evidence only shows natural selection explaining rather small changes, and I see profound difficulties in thinking that it explains much more than trivial changes. It is fine by me if common descent is indeed true, and there is some sort of designed program to power changes over time (i.e., evolution). And I think things like pseudogenes are strong arguments for common descent. So again I'm an 'evolutionist' in that sense." Dembski: "Right now I'm inclined toward a preprogrammed form of evolution in which life evolves teleologically (humanity being the end of the evolutionary process)." The sympathetic philosopher Alvin Plantinga: "I'm a creationist in the sense that I think God created the heavens and the earth and all that they contain, but I'm not prepared to say he didn't do it by way of evolution. He could have done so by orchestrating the appearance of the right mutations at the right time, etc. So I'm not wedded to special creation, tho I do think the special creation of life, and maybe (in some way or other) of human beings is more probable than not. But even without special creation it would still be the case (so I say) that God created human beings in his own image." Even Johnson is not a flat-out, take-no-hostages opponent of evolution: "I agree . . . that breeding groups that became isolated on an island often vary from mainland species as a result of interbreeding, mutation, and selection. This is change within the limits of a pre-existing type, and not necessarily the means by which the types came into existence in the first place. At a more general level, the pattern of relationships among plants and animals suggests that they may have been produced by some process of development from some common source."[27]

Special interventions up front; evolution in the background. Of course one can point out that traditional creationists allow something like this. Price allowed change to occur after the flood, to obviate the space limitations of the Ark. The worry among creationists has never been transmutation as such but more the overall picture that it represents. And the same is true of the ID enthusiasts. What is driving them is their opposition to naturalism as a philosophy. Every person promoting this position admits frankly that he sees Darwinism as the epitome of a "naturalistic" system which puts all down to blind forces working on inert matter. And every one of these people loathes naturalism because he sees it as a brief stop on the route to atheism. All feel that, as is shown only too well by the frenzied polemics of Richard Dawkins and others, Darwinian evolution equals naturalism equals no Christian God.

Naturalism is a *metaphysical* doctrine, which means simply that it states a particular view of what is ultimately real and unreal. According to naturalism, what is ultimately real is nature, which consists of the fundamental particles that make up what we call matter and energy, together with the natural laws that govern how those particles behave. Nature itself is ultimately all there is, at least as far as we are concerned. To put it another way, nature is a permanently closed system of material causes and effects that can never be influenced by anything outside of itself—by God, for example. To speak of something as "supernatural" is therefore to imply that it is imaginary, and belief in powerful imaginary entities is known as superstition.[28]

One might protest that the naturalism of the scientist is only an assumption. Science employs a methodological strategy, a way of approaching knowledge of the natural world; it does not make a metaphysical commitment to a preconceived way the world must work. Someone like Simon Conway Morris might say that as a scientist he allows no explanation beyond the natural, though as a Christian he believes in a realm beyond the natural world. Johnson and his associates will have nothing of this. Such a compromise never works and is always the thin edge of a wedge that leads to atheism.

Plantinga goes so far as to argue that naturalism is self-defeating even as a strategy for approaching knowledge, for it makes impossible any genuine knowledge at all. If naturalism is true, then Darwinism is plausible—in fact, it is the best game in town. But Darwinism is not interested in our getting to the truth about reality; it is concerned only with survival and reproduction. And as a reproductive strategy, believing that this world is all there is might be advantageous. Advantageous, but not necessarily true. Conversely, believing in a Christian God might be advantageous, but it also might not be true—just an epiphenomenon of selfish genes. Naturalism leaves us with no way to know, one way or another, the truth about reality, in this world or in worlds beyond. A belief in the power of natural selection leads logically to acceptance of a perpetual state of deception.[29]

Does rejection of naturalism and acceptance of the truth of Genesis necessarily entail a *literal* reading of the creation story? On this question, no unanimity exists among the most prominent intelligent design theorists. At one extreme, Johnson is prepared to hypothesize that the earliest men (Methuselah and company) may have indeed lived much longer than people do today. The philosopher of science Paul Nelson comes from a family tradition of several generations who were sympathetic to young-earth creationism, as is he. "I am a 'theological' young-earth creationist, which means that I think the bible makes much more sense when read in that light."[30] Others stay away from this kind of speculation. But within basic limits (everyone wants something special for humans, spiritual if not physical), literal interpretations are not the big issue for these men. Many of the leading intelligent design theorists do not belong to churches where one would expect a neo-Adventist reading of Genesis. Behe is a Roman Catholic. Dembski went to Princeton Theological Seminary and was a Calvinist before recently moving in the direction of Eastern Orthodox Christianity. Plantinga is a Calvinist. Johnson is an evangelical Presbyterian.

Obviously, with these diverse attitudes and religious commitments, we should not expect to find a consistent line of ardent premillennialism. Some ID supporters do fall into this camp. Nelson is a deeply committed premillennialist. He thinks, for example, that the founding of Israel was highly significant; his father twice tried to move the whole family there. "I was sent to Jewish nursery schools as a child, and learned Hebrew as a boy, rising early in the morning to practice grammar and vocabulary." His brother-in-law is a Jew who converted to Christianity and returned to the Promised Land to live and to prepare for things to come: "Great shakings and darkness are descending on Planet Earth, but they will be overshadowed by even more amazing displays of God's power and light. Ever the long-term strategist, YHVH is raising up a mighty army of cutting-edge Jewish end-time warriors." Another intelligent design theorist, the philosopher Stephen Meyer, eschews any formal system of eschatology but allows that he is "intrigued with the idea that the modern state of Israel might probably constitute a realization of ancient prophetic expectations." Continuing: "In fact I think a strong case can be made for this."[31]

From others we get a form of (at most) amillennialism. Behe: "To tell the truth, I look at all this 'post-millennial', 'pre-millennial', etc., as Protestant-talk. Catholics don't think like that." Dembski: "Within Eastern Orthodoxy, the view is that we are now experiencing Christ's millennial reign. I have a certain affinity to this view." Plantinga: "I'd go with Augustine here: I don't take any position on millennialism, and don't think it's important re Christian belief." Johnson: "On eschatology, I am content to 'wait and see what happens.' We hardly ever mention the rapture or Israel in the Intelligent Design movement. I do say that 'it is unwise to try to force God's hand.'"[32]

Some ID supporters may be more amillennial than premillennial, but they all strongly oppose postmillennial hopes and programs for action. Paul Nelson is typical: "Any human attempt to bring 'heaven on earth' is doomed to misery . . . Reading Solzhenitsyn and Orwell as a teenager made me a lifelong doubter of such schemes."[33] Meyer has written against proposals for state-backed universal health care and returns repeatedly to condemn unlimited access to abortion procedures. Johnson is strongly against ideas of progress. "The Christian philosophy that was overthrown in the 1960s was an easy target because it had become identified with American culture and with worldly ideas like human perfectibility and the inevitability of progress, which are actually profoundly un-Christian."[34] Along with the others, he would urge on us personal morality rather than schemes for reformation of society. Evolution is tied to a philosophy of action and attitude that leads straight to all the woes of modern life—homosexuality, crossdressing, promiscuity combined with fornication and adultery, easy divorce, abortion, lack of respect for duly constituted authority, socialism, and more (single mothers, state welfare, absent fathers, and jail time for teenage sons).

As part of their rejection of methodological naturalism, ID theorists reject the label of "theistic evolutionist," meaning someone in the Asa Gray or Teilhard de Chardin mold who thinks that God acts through evolution in a creative fashion. Apparently, theistic evolutionists are methodological naturalists. "Theistic evolutionists generally accept the entire Darwinian scientific picture, but say that God was invisible and undetectably behind it. For them God's participation is known only by faith and not by anything detectable by scientific investigation."[35] ID theorists insist that one can infer God's existence from the evidence. Hence, even though one cannot properly, without qualification, simply include all of the ID theorists in the creationist camp, one has good reason to see the group as part of this ongoing tradition. It is an American tradition that goes flatly against the secular religion of Darwinism, and does not have a whole lot of time for the proposals of more liberal Christians. As always, the battle is not simply one of fact and truth. It is rather a struggle for the hearts and souls of people, with deep implications for the ways in which we live our lives and regulate our conduct. It is a religious or metaphysical battle, not simply a dispute about scientific theory.

Conclusion

In the sixteenth century Nicholas Copernicus
told us the earth was a ball and, what was worse,
was not the center of the universe.
"Well and so," we wanted to know,
"where does that leave us in the scheme of things?"

Wherever it left us,
we were just about learning to live with it,
when three centuries later Charles Darwin
grabbed our attention with the news
that we were cousins to the kangaroos.
"And so," we wanted to know,
"where does that leave us in the scheme of things?"

*T*his question, posed by the Arkansas poet Miller Williams, is the one with which we have been grappling throughout this book. After Darwin, where does that leave us in the scheme of things? Let us look back, and then see what lessons we can extract from this history.

The starting point to our story was a crisis of faith during the Enlightenment. No longer did the old religious verities convince. There were problems with revealed religion (how does one choose between the options, or indeed between Christianity and foreign faiths?), and there were problems with natural theology (how does one explain pain and suffering if God is as good as he is supposed to be?). This crisis did not come about all at once; the water just dripped away at the stone. Science was certainly a factor in the crisis of faith, but it was not the only one. At least as big a cause was changing social conditions. In England, as the population became increasingly urban, Anglican Christianity could no longer maintain the emotional hold on the masses that it once had enjoyed. And in America following the Revolution, new ideas were needed for a new land with a new government.

Whatever the causes of a loss of faith, Matthew Arnold captured its cadence:

> The Sea of Faith
> Was once, too, at the full, and round earth's shore
> Lay like the folds of a bright girdle furl'd.
> But now I only hear
> Its melancholy, long, withdrawing roar,
> Retreating, to the breath
> Of the night-wind, down the vast edges drear
> And naked shingles of the world.

Darwinian evolution was not, in itself, a direct response to this crisis of faith. History did not start abruptly in 1859. If (to pick a date) we say that the crisis started back around 1700, then equally we can say that reactions to it started back around that time as well. However important evolution may be to the story of Christian faith, it was only one among many threads, and not necessarily the most significant one. But with these qualifications noted, a fairly clear picture of an ideological struggle starts to emerge, with evolution near the center. Although there were those who tried to walk the middle path, retaining the old faith while accepting new scientific knowledge, two polarized, innovative strategies emerged. One was to make reason and progress the central focus of one's worldview; the other was to make emotion and faith the central focus. Let us elaborate, taking these strategies in turn, showing how evolution was part of both pictures, and how they were guaranteed to clash.[1]

On the side of reason, long before Charles Darwin the rise of science led to beliefs in progress and opposition to Christian Providentialism. The success of lawbound explanations put pressure on appeals to miracle, and the success of technology made people think it was possible to do things without God's help. Related philosophical discussions cast the old truths in doubt. Some people tried to stay within an explicitly Christian framework, while others moved to a more deistic and later more naturalistic (and agnostic or even atheistic) picture of the universe. But for everyone on the side of reason, progress and evolution were inextricably entwined. This picture was infused by a postmillennial eschatology, which saw humans as working toward a brighter tomorrow. Even those who wanted to retain some vestiges (or more) of Christian belief thought that the Second Coming would occur sometime in the indefinite future, but only after we had (in William Blake's words) "built Jerusalem, in England's green and pleasant land." It is no surprise that this much-loved hymn became the anthem of the socialist-leaning British Labour Party.

This progressive picture was never intended to be purely scientific. The issue was not the merits of science as such but rather what one was to do in the face of a loss of traditional faith and its supports. Evolutionists responded by making a kind of religion out of their science, stressing its supposedly progressive nature as a substitute for traditional beliefs. Some of the most respected men of the day—including Charles Darwin's own grandfather—went beyond trying to devise a theory of evolution and endorsed a philosophy of evolutionism, with all the social improvements it entailed. Charles Darwin inherited this tradition along with the family china; and for all that he himself may have wanted to produce a straight, value-free, scientific theory, his work and thinking did impinge on the traditional religious picture, changing if not eliminating much of its supportive theology.

Darwin showed that blind law can explain the diversity of organisms, thereby making miracles even less plausible and putting another nail in the coffin of the argument from design. More importantly, his supporters made his thinking the keystone of a now-respectable evolutionary substitute for Christianity (or, for some, a Christianity supplement). And so the story has continued. From Thomas Henry Huxley and Herbert Spencer down to William Hamilton and Edward O. Wilson, evolution took on the trappings of a religious faith. It offered a story of the origins of life and a meaning to existence. The Alpha and Omega. A religion of evolutionary naturalism that was thoroughly postmillennialist, in the sense that it pointed to a brighter tomorrow if only we do what is demanded of us.

On the other side, we have those who countered the doubts and arguments of the Enlightenment by favoring emotion over reason. A commitment, through faith, to Jesus as the Lord and Savior who died on the Cross for our sins was the necessary and sufficient condition for salvation. These evangelicals turned away from modern philosophy and religion in their various guises, along with materialism and the other implications of modern science. Rejecting progress, they reaffirmed their commitment to Providence. Especially in America, the Bible became the central text of this movement; and only the pure and untutored heart held the key to understanding.

This evangelicalism was as innovative and creative, in its own way, as evolutionism. While evolutionists were drawing on venerable postmillennial Christian themes of progress and perfectibility, the evangelical creationists were drawing increasingly on premillennial themes, which made them, especially as the nineteenth century got under way, no more like traditional Christians than were the social Darwinians.

Saint Augustine would have been as mystified by (and opposed to) the teaching of John Nelson Darby as by that of Herbert Spencer. Evolutionism and evangelicalism were, both, new answers to a new problem: the threatened loss of faith. For the evangelicals, the work and legacy of Charles Darwin became a focal point of opposition— the alternative religion that must be fought on every front and by any means. These believers did not resent evolutionary change as such— almost every creationist allows some change across species. What they found objectionable was the context in which such change was presented—a self-sufficient context of naturalism and unbroken law, from first to last. For evangelicals, a biblically based story of origins had to be articulated and defended (creationism), a way of life based on this picture had to be endorsed (some version of Puritanism), and a future stemming from this picture had to be proposed (premillennialism).

And do note just how fundamental was this underlying eschatology. What the evangelicals opposed was a naturalistic, evolutionary world leading to a brighter tomorrow. If need be, the opponents of this progressive picture were prepared to throw in dollops of evolutionary change to protect their own resolutely anti-postmillennialist picture—gaps in the fossil record were not what kept these evangelicals awake at night. What disturbed their sleep was opposing visions of the future and beliefs about the right way to live in this world, today.

So much for the history. Now, what does this teach us? I will make five points.

Rival Religions

Why is there a clash? By the beginning of the twentieth century, evolutionism and creationism were competing for space in the hearts and minds of regular folk. It was not a science-versus-religion conflict but a religion-versus-religion conflict—always the bitterest kind. Darwinians did not have to become secular theologians, but many did. Evolution did not necessarily entail evolutionism, but many evolutionists made the move. Likewise a Christian did not have to become a creationist (or anti-Darwinian), but many did. Creation did not necessarily entail creationism, but many creationists made the move. Hence, the controversy. And here is the reason for the controversy:

> And we are here as on a darkling plain
> Swept with confused alarms of struggle and flight,
> Where ignorant armies clash by night.

Quarrels are always intense when they happen within the family. Siblings share the same background and experiences, the same hopes and challenges. Indeed, their growth comes from defining themselves in contrast with one another. They accept the same terms of engagement, so when disputes arise, they fight head on, rather than flailing around against an enemy they don't understand. Likewise with evolutionism and creationism. They have the same parent (a crisis of faith) and they have grown and developed in major part by defining themselves in opposition to each other. They accept the same set of issues as being crucial. But always, they have different solutions. Evolutionists are postmillennialists, thinking that we humans can improve our lot and bring about heaven on earth. Creationists are anti-postmillennialists, whether they be traditional premillennialists or modern amillennialists (some of the intelligent design theorists). They are against massive government-backed programs for societal change and would have us concentrate rather on immediate and personal salvation.

A large survey of 4,000 Americans in spring 2004, sponsored by the Pew Forum on Religion and Public Life, defines "traditionalist"

Christians as those who tend to literal readings of the Bible, non-metaphorical understandings of basic claims about God and the afterlife, and nonbelief in evolution—in other words, they map those whom I have labeled creationists. They cluster around premillennialist-type beliefs. At the other end of the pole on such issues as the Bible, God, and evolution are "modernist" Christians. They cluster around postmillennialist-type beliefs. If you factor in affiliations—people like Southern Baptists who belong to evangelical churches (26.3 percent of the country), as opposed to people like Episcopalians who belong to what were formerly (but no more, as these numbers show) mainline churches (16 percent)—the traditionalist-modernist divide becomes even stronger. Moreover (and justifying the Vatican's ongoing worries), in American Catholicism, too, the seeds of the Enlightenment bear more fruit than the seeds of the Reformation. Catholics (who make up 17.5 percent of the U.S. population) are no less divided between traditionalists and modernists. And those people without a formal affiliation with a church (16.0 percent of the U.S. population), especially agnostics and atheists (3.2 percent), reinforce the picture, subscribing strongly to a postmillennialist philosophy, even where there is no supporting theology.

In a comparison between social welfare issues and so-called cultural issues such as abortion and crime, which count as most pressing issues to the people surveyed? Among evangelicals, the judgment was 27 percent welfare to 40 percent culture. In contrast, modernist mainliners had ratings of 46 percent welfare and 11 percent culture. Secular thinkers were virtually the same as modernist mainliners. All Catholics tended to place welfare higher, but modernists came in at 50 percent (with a mere 6 percent rating culture more highly), as against traditionalists, who came in at 39 percent (with 25 percent rating culture more highly).

Turning specifically to abortion, only 9 percent of traditional evangelicals think it should be a woman's choice entirely, whereas 62 percent of modernist mainliners think this way, and the same is true of the secular thinkers. Notwithstanding official Church policy, a similar divide exists between traditionalist and modernist Catholics: 17 percent and 54 percent, respectively. Although all Christians tend to be more conservative than nonbelievers on such issues as the death penalty and homosexual rights, the pattern is much the same. For instance, on gay marriage, only 3 percent of traditional evangelicals support it, a minority of 38 percent of mainline modernists support it, but an overwhelming 72 percent of atheists and agnostics support it. Traditional and modernist Catholics split 11 percent to 51 percent.

Finally, given the significance of the restoration of Israel in premillennialist thinking, whereas 64 percent of traditionalist evangelicals support Israel over the claims of Palestinians, only 22 percent of modernist mainliners and 15 percent of nonbelievers do. Catholics split at 43 percent traditionalists, 23 percent modernists. Jews are the biggest supporters of Israel (at 75 percent), but since 1992 Jewish support has dropped somewhat, while the support of traditionalist Christians keeps moving up.

These numbers point to a stark theological divide—a divide in which belief in an all-encompassing evolution is a central defining point, and from which moral prescriptions flow. And so the hostilities spill out, with the two sides sounding like grotesque parodies of each other, in their obsession with similar topics. Consider first W. D. Hamilton on family values and practices:

> One of the ways in which I think backing plus curbing of the hypocrisies of individualism will come about will be through a greater measure of *family* responsibility that political parties will see it as a necessary measure to impose: as example, if a family wants to keep a particular vegetable baby alive, the family must

pay for it. Similarly, if a church objects to the alternative—letting the baby die—extra taxes to pay for the baby's special care will be required from specifically that church's coffers, backing its beliefs. None will be demanded from another church that agrees the baby should die. In general along such lines, it will be a great step in the equitable running of modern society if a sincerity tax comes to be imposed on all propaganda—what you say you believe in you must show you believe in through hard cash and sacrifice; as an example again, there should be no option but that your child attends the idealistic comprehensive school you say you believe in.

But why not start to discuss and make decisions and try experiences now? Just as I believe Kosovars should be allowed to practice Islam in Kosovo, in the heart of Christian Europe, if they want to (including even practice the purdah and female circumcision if they want that) but not including any right of unlimited increase (or at least providing very strong disincentive, in contrast to the recent supposedly liberal and yet embittered spirit of toleration), so I believe that other self-defined yet quite different groups of idealists should be allowed to practice a religion that includes parent-decided selective infanticide, provided this is done under good safeguards against cruelty. Through free choice in idealism this will become real, effective group selection on the cultural plane, hopefully no longer contaminated from its warlike predecessor out of biology.[2]

This extreme view comes from a figure who towered in the world of evolution. In the words of Richard Dawkins, speaking of Hamilton: "Those of us who wish we had met Charles Darwin can console ourselves: we may have met the nearest equivalent that the late twentieth century had to offer."[3] Turn now to the other side—the opinions of Phillip Johnson:

A responsible society is based first and foremost on responsible parents who fulfill their obligations to each other and to their children. Probably the most important thing that most adults do is to prepare the next generation for the joys and responsibilities of life. To do this they must ensure to the best of their ability that their children are born healthy. Following birth, children must be nurtured and educated in moral behavior by loving parents, preferably *two* parents. That is one reason it is important for lovers to regard marriage as a sacred bond, rather than as a contractual arrangement to be terminated at the convenience of either party. That is also why mothers in a rational society regard their children, born and unborn, as a sacred trust rather than primarily as an encumbrance that men impose on women in order to make them unhappy and impede their pursuit of wealth, power and pleasure. Similarly, fathers in a rational society regard their offspring from the beginning of pregnancy as their own flesh, so that they become enthusiastic providers and conurturers rather than the unwilling objects of child-support orders.[4]

Among anti-Darwinians, Johnson holds the respect and affection that Hamilton's memory holds among Darwinians. These two positions on family obligations reflect and distort each other like crooked mirrors at the fairground.

Sensible Strategies

Thomas Kuhn distinguished between puzzles and problems. Puzzles are challenges that have solutions. That is why we speak of crossword *puzzles.* Problems are conflicts that may well not have solutions. That is why we speak of the mind-body *problem.* In the real world, we know too well about problems. Northern Ireland is a problem. The Middle

East is another problem. Solutions or not, problems cannot be ignored. One sensible strategy, when faced with a conflict, is based on
the old Arab adage: "The enemy of my enemy is my friend." In World
War II, Britain and America had little liking for the Soviet Union but
recognized that a nation which was fighting so ferociously against the
German army should be encouraged and assisted wherever possible.

Creationists do a good job of making friends with the enemies of
their enemy. There are clearly massive differences between premillennial dispensationalists like Henry Morris or Tim LaHaye and ID enthusiasts like Michael Behe and William Dembski. They recognize
this themselves. The former want strict, young-earth creationism,
with six literal days of creation, a universal flood, a premillennial
rapture, and the rest. The latter pull back from most of this, finding it
as crude and inessential for proper Christian faith as does a Christian
evolutionist like John Haught. Nevertheless, although differences do
sometimes surface—John D. Morris, the son and successor of Henry
Morris, complains that ID "doesn't go far enough"—members of this
side recognize shared aims and the necessity to forge alliances.[5]

The dispensationalists have made friends with the intelligent design faction because "the cultural elites at present fear the ID movement more than they fear biblical creationism." In appreciation, the
dispensationalists give the ID proponents favorable treatment in their
publications; for example, Dembski was praised in a women's magazine run by the wife of Tim LaHaye. All to a future end: "If the science establishment can be forced to acknowledge the scientific case for
intelligent design, theism will become part of the 'post-Christian' cultural air. In that philosophical environment, a new set of options will
be open for people, one of which will be biblical creation."[6]

If only evolutionists could get their act together like this. Admittedly, some Christian evolutionists reach out to modern science—Simon Conway Morris most obviously. On the other side, admittedly

some nonbelieving evolutionists are willing to reach out to Christians and embrace their hopes and desires. Perhaps because of his upbringing in the Bible Belt, or perhaps because he sees religion as being part of the essential human condition, Edward O. Wilson has always made efforts to speak with Christians, despite his own clearly stated nonbelief. Recently, in his work on ecology and biodiversity, Wilson has referred to Holmes Rolston with approval as an appreciated and respected authority. In *The Future of Life*, Wilson quotes and endorses an anecdote by Rolston: "For years trailside signs at a subalpine campground in the Rocky Mountains he occasionally visited read, 'Please leave the flowers for others to enjoy.' When the wooden signs began to erode and flake, they were replaced by new ones that read, 'Let the flowers live!'"[7] Both men think life has value for its own sake, independent of whatever enjoyment it brings to humans.

Generally, however, the misunderstandings and hostilities among evolutionists are frightening. Many if not most Christian evolutionists, following in the tradition of Teilhard de Chardin, reject twentieth-century biology or strive desperately to supplement or replace Darwinism. Few are quite as candid as Keith Ward, but one finds it hard not to suspect similar motives at work among people like Rolston. In part, this stems from an understandable dislike of the strident and intentionally hurtful atheism of Dawkins and his kind. Who would want to agree with such a person, even about science? In part, discomfort with modern science comes because Christians find Darwinism itself too challenging. But at least the Christian evolutionists strike a civil tone in their critiques.

The same cannot be said for the other side. The sentiments of Dawkins and Coyne are but the tip of a very chilling iceberg. The well-known historian of evolutionary biology William Provine is a paradigm: "You have to check your brains at the church-house door if you take modern evolutionary biology seriously."[8] Even those who claim to be religion-friendly tend not to be, on closer examination. Despite Gould's lip service to "Magisteria," as soon as a Christian wants to make any existential claims—for instance, that God the Creator sent his son to earth to save us from our sins so that we can enjoy salvation and an afterlife—Gould dismisses the claim (and by implication the person) as silly or stupid.

The point is not that evolutionists must all accept Christ, or Christians must all accept Darwinism, for understanding to flourish. Roosevelt and Churchill did not become communists while cooperating with Stalin. The point is rather that the two opposing sides within the evolutionary camp should spend more time focusing on the huge amount of intellectual and emotional overlap in their respective positions. The overlap starts with a common commitment, among many Christians and evolutionists, to a postmillennial philosophy, and it moves on to scientific details like the way that both Dawkins and Rolston appeal to ratchetlike effects in arguing that evolution is progressive.

And given the threat that creationists pose to evolutionists of all kinds, it behooves evolutionists especially to start thinking about working together with Christian evolutionists, rather than apart. For a start, atheists like Dawkins and Coyne might consider taking a serious look at contemporary Christian theology (or the theology of other faiths, for that matter), rather than simply parroting the simplistic, schoolboy travesties of religion on which their critiques are founded. Conversely, Christians like Ward and Rolston might be encouraged to dig more deeply into modern, professional evolutionary biology and to start to get some understanding of its strengths and triumphs before they cast around for alternatives like self-organization. Even if the latter turns out to be important, this importance will need to be understood in the light of Darwinism.

Evolution as Science

Now to move to a third point: the distinction between evolution as a scientific theory and evolutionism as a religious commitment. Please do not misread me as saying that evolution in all forms is nothing but a secular religion. Our history confirms that slowly and gradually but definitely evolutionary thinking dragged itself up until at some point in the 1930s a mature, professional kind of evolutionary theorizing emerged. Neo-Darwinism (the synthetic theory of evolution) became a genuine science, one that gives us truths about the way the real world works. From evolutionism emerged a serious theory of evolution.

It strove to be culture-value-free, but the crucial issue is not the presence or absence of values as such but rather the kinds of values that guide and inform science. Pulling together points made in earlier chapters, we can see that a mature science avoids or eliminates cultural values (like "Europeans are better than other races" or "Christianity is the best religion"), yet is informed by what philosophers speak of as "epistemic norms" or values—the sorts of rules that physicists and chemists, the most successful of scientists, use to guide them in their work. The ultimate value is understanding the world in which we live. Knowledge is in itself taken to be a good thing—a value—and various norms or rules are generally accepted by scientists (and philosophers of science) as being important in attaining this understanding. Included is predictive ability. A theory that enables one to make accurate forecasts about what will be found is thought to be more than a mere figment of the theorist's imagination. Highly desirable also is unificatory power, or what William Whewell (identifying it as the focus of a *vera causa*) called consilience. A theory that binds together various disparate areas of knowledge is thought to have more power and worth than a theory that explains only one separate piece. Coherence within and consistency without is important too. No one wants a theory that makes contradictory claims, or one that if true then simply makes everything else, however apparently well-founded, false. And fertility is a final important consideration. A theory that points the way to a fruitful research program—one that raises a broad range of questions which can be successfully answered through careful observation and experimentation—is much more valuable than a theory that is impotent. None of this respect for epistemic norms is a substitute for going out and observing the natural world, or doing experiments in the laboratory; grappling with the empirical world is the name of the game for the professional scientist. Rather, these epistemic rules, or scientific rules, are what guide scientists as they peer into the nature of reality.

These rules are not written on tablets of stone, like Platonic forms waiting to be intuited by the truly rational being. If anything, our history suggests that an element (at least) of pragmatism enters in, since they have often been adopted for social rather than intellectual reasons. Nevertheless, these epistemic rules or norms, which are accepted as the determinants of the best kind of science, are used by and govern evolutionary theorizing today. Prediction, consilience, consistency, and fertility have been important for W. D. Hamilton, E. O. Wilson, and all the great evolutionary biologists of our day. Indeed, these biologists' work is prized precisely because it is of such high epistemic quality. Kin selection, for example, has given research projects to a whole generation of students. The same is true of evo-devo. Stephen Jay Gould's theory of punctuated equilibrium, on the other hand, has been controversial precisely because of doubt about its ability to lead to exciting new predictions and research, or to cohere with what we know at the microevolutionary level.

Even if epistemic rules have a pragmatic element, this does not mean that all is relative, or ultimately a matter of cultural value. Many historians and sociologists of science have embraced some form of social constructivism, seeing science as a reflection of culture and cultural values—as merely endorsing the norms and practices of the society of the day. Our history does not justify such a broad conclusion. It is true, however, that evolutionary theory past and present reflects culture to a large extent. It arrived on the scene as a rival answer to Christian questions about origins, and then it proceeded to appropriate the ideas and concepts of Christianity in order to do much of its explaining. For a Darwinian even today, final cause is as important a notion as it was for the most died-in-the-wool natural theologian. Richard Dawkins has joked, with much justification, that he is somewhere to the right of Archdeacon Paley on these matters.

Final cause is the thin end of a very thick wedge that takes us into the realm of metaphor. In thinking of the heart as a pump, for example, one sees (as Aristotle noted) organic features as if they were designed. More broadly, science, including evolutionary science, is drenched in metaphor—the struggle for existence, natural selection, balance of nature, division of labor, adaptive landscape, selfish gene, arms race. This is not necessarily a flaw. Without metaphors, science could never exhibit epistemic virtues. The metaphor of natural selection enabled Darwin to build his great consilience. The metaphor of an adaptive landscape let the synthetic theorists in the 1930s blend Darwinian selection fruitfully with Mendelian genetics. All metaphors are deeply cultural. The division of labor is right out of Adam Smith's economics of the eighteenth century, and arms races are part of the dreadful heritage of the twentieth century.

Yet, the use of metaphors does not necessarily imply endorsement of the *values* that they might be thought to represent. A modern biologist at work, like Edward O. Wilson studying the so-called caste systems of ants, is not necessarily thinking that a division of labor in our human society is a good thing, even though this metaphor guides him toward understanding why the soldier caste of ants never cares for the young, or the nursery attendants never go foraging for leaves. Wilson could well think that, in the human world, a division of labor is psychologically deadening and counterproductive, but ants are not humans and so do not suffer from boredom. Likewise a Quaker biologist, a pacifist, might abhor arms races in the real world and yet find the notion very useful as he or she traces the evolution of a snail and its shell through various strata. If understanding the interactions between snails and predators cries out for an arms-race analysis, why not use it? Snails are not Russians and predators are not Americans.

Intelligent Design

Evolutionism may be a religion, secular or otherwise. This does not imply that all evolutionary theorizing is a religion. To the contrary. Good reasons exist to think that modern evolutionary biology is genuine science. So why not extend the same courtesy to the other side? Why not allow that, although full-blooded creationism is a religion, intelligent design theory is potentially if not actually real science? Sauce for the goose is sauce for the gander. This is not a silly question, but the case must be made on its merits, and there's the rub. If we judge by the epistemic criteria introduced in the last section, we find no empirical or conceptual reason whatsoever to think of intelligent design theory as genuine science.

On the important criterion of fertility—does the science provide work for the new generation?—ID theory flops dreadfully. Dembski and other enthusiasts realize that it should do this: "The world contains events, objects, and structures which exhaust the explanatory resources of undirected natural causes, and which can be adequately explained only by recourse to intelligent causes. Scientists are now in a position to demonstrate this rigorously. Thus what has been a long-standing philosophical intuition is now being cashed out as a scientific research program."[9] But there are no results. And there are no new predictions leading to new and unexpected discoveries.

Several creationists have done genuine science in the past. Duane Gish once coauthored a paper published in the *Proceedings of the National Academy of Sciences.* But the work being done today in intelligent design is not of that caliber. In 1997 George Gilchrist reported the results of his search through five huge computerized databases (BIOSIS, the Expanded Academic Index, the Life Sciences Collection, Medline, and the Science Citation Index). Among hundreds of thousands of publications, he found only thirty-seven references to intelligent design as a biological theory, and none of them reported scientific research based on the theory. Later searches extending from 1997 through October 2001 produced the same result: no scientific research into intelligent design.[10]

We might go at things in a slightly different way. Instead of offering an empirical discussion—searching and counting publications— we might try to decide matters by reference to our use of the word "science." Specifically, we might simply argue that ID theory could not possibly be genuine science because, by definition, genuine science makes no reference to miracles or other extranatural interpositions into the course of nature. Dembski would probably object that ID theory does not necessarily require miracles—the essence of the theory is intelligent intervention. But he is not entirely candid on this matter. "The crucial breakthrough of the intelligent design movement," he has written, "has been to show that this great theological truth—that God acts in the world by dispersing information—also has scientific content." So, we seem to be back to miracles after all.[11]

Nevertheless, a disinterested observer might feel that this is all a little bit quick and slick. Even if we slide over Dembski's ambiguities, ID theory cannot be dismissed just because science since the Scientific

Revolution has not allowed miracles as explanatory causes. Perhaps the time has come to reverse the trend, if indeed miracles can be shown to exist. Perhaps science should allow miracles, and the miracles of ID theory are a place to start. Otherwise, science, faced with irreducibly complex phenomena, must remain incomplete. The philosopher Alvin Plantinga suggests that if you do not like to use the word "science" here, then replace it with a new term like "Augustinian science" (because presumably this is the kind of science that the great philosopher would recognize). The point is that you should not just refer to a lexical definition of "science" and think that this proves something about the way the world really works.

Granting that reference to definition certainly will not do the job of ruling out the worth of ID theory, let's go back to the purpose of science and to the epistemic virtues of respected fields like physics and chemistry. Plantinga himself agrees that recourse to miracles is a "science stopper." But as a philosopher and as a Christian, he is happy to accept this.

> The claim that God has directly created life, for example may be a science stopper; it does not follow that God *did not* directly create life. Obviously we have no guarantee that God has done everything by way of employing secondary causes, or in such a way as to encourage further scientific inquiry, or for our convenience as scientists, or for the benefit of the National Science Foundation. Clearly we cannot sensibly insist in advance that whatever we are confronted with is to be explained in terms of something *else* God did; he must have done *some* things directly. It would be worth knowing, if possible, which things he *did* do directly; to know this would be an important part of a serious and profound knowledge of the universe.[12]

Scientists (including scientists who are Christians) respond that science has made its greatest advances precisely by refusing to allow science stoppers. If you invoke miracles every time you run into something you cannot readily explain, you will get nowhere. Scientists make the discoveries they do because they refuse to give up and put unexplained phenomena down to non-natural causes. The faith of scientists in methodological naturalism is not a function of some unacknowledged atheistic program but simply a reflection of the fact that toughing it out has worked in the past. It is sensible to assume that this strategy will work in the present and future also.

Whether or not this response is justified—and many evolutionists think that already we know much about the natural origins of the supposedly irreducibly complex phenomena that Behe highlights— the fact is that science stoppers do just that. They stop science.[13] They stop the quest to understand nature according to the constraints of epistemic rules and norms. And what sticks in the craw of scientists is that the only grounds for doing so are extrascientific—people of faith just want to bring in religion. So even if Plantinga is right, and even if ID theory does give us "an important part of a serious and profound knowledge of the universe," that knowledge is not scientific knowledge. It cannot replace the understanding of life gained through contemporary evolutionary theory.

Evolution or Evolutionism?

We come now to the fifth and final point. The unsentimental evolutionist will complain that too much ado is being made about too much nothing. Evolution today is a genuine scientific discipline. Leave it at that. Evolutionism—making evolution into something more than science—is the cause of the trouble. Darwinian evolutionary theory, or some version thereof, is no more relevant to the problem of creationism than, say, quantum mechanics or ideal gas theory. Perhaps a political battle must be fought against the creationists, but evolution does not have a dog in that fight. To think otherwise is to loiter in the past or to confuse the professional and public domains. And let us not forget that creationists themselves often accept some measure of evolution!

If only things were this simple. First, science or not, Darwinian evolutionary theory does impinge on religious thinking. Richard Dawkins is absolutely correct when he speaks against those, like John Henry Newman, who claim that science and religion are about different things and hence can never conflict. Granted, a truly traditional Christianity does not conflict with science (including Darwinism), in the sense that evolution as such does not oppose religion as such, even though secular evolutionism may. But if the notion of "religion" is extended to every belief system that people have held as giving insight into the sacred, then there obviously can be and is conflict with science. Mormons believe things about native Americans that are in flat opposition to modern anthropology.

In the case at hand, Darwinian theory goes against Genesis, taken literally. Darwinians may be right on this point, and Christians who refuse a metaphorical reading may be wrong on this point, but that is not the issue. The issue is that Darwin's theory considered as science is pertinent to the conflict. Likewise, Darwinism makes the argument from design as a proof for God's existence less than compelling. As Dawkins has said, after Darwin—and only after Darwin—has it been possible to be an intellectually fulfilled atheist. Therefore, Darwinians, as members of society who are often in teaching positions and whose research costs are generally supported by others, including others who do not like their theory, have a special obligation to think

about the broader implications of their work. Not to quit doing science or stop being evolutionists but to think about ways in which even
they might contribute to tackling the evolution-creation dispute.
These are not trivial, surface questions. Phillip Johnson is not
entirely wrong when he links methodological and metaphysical naturalism. One can be a methodological naturalist and a metaphysical
supernaturalist at the same time, and, contrary to Johnson, this is not
a stupid or inconsistent position. Many serious Christians have gone
this route in the past century, including Ronald Fisher, Theodosius
Dobzhansky, and Simon Conway Morris. Whether or not one accepts all that they say about the nature of life's history, they are in no
sense scientifically inadequate or religiously insincere in arguing as
they do. It is simply not the case that people take up evolution in
the morning, and become atheists as an encore in the afternoon. Detailed study of the obituaries of nineteenth- and early-twentieth-
century secularists shows that rarely if ever did people become nonbelievers through reading Darwin or Huxley. Much more influential
were the inconsistencies in the Bible and the arguments of radical
tracts, like Thomas Paine's *The Age of Reason*.[14] But, this said, many people have shown a tendency to slide from methodological naturalism toward metaphysical naturalism, as Charles Darwin's own history
shows. Once you take God out of the equation for running the natural world, it's a small step, for some, to arguing that he is not there at
all. And many people are clearly inclined to do just this. Particularly
on the question of origins, even though many evolutionists may themselves be willing to make the slide, they should not be surprised when
others, seeing a slippery slope from methodological naturalism to
metaphysical naturalism, stop themselves at the top of the hill. Evolutionists who go the whole route, sliding down to an explicitly atheistic
metaphysical naturalism, might do well to ask whether they can justify

ending up where they do. For those who answer yes, are they justified in arguing that others who do not follow are simply naive or inadequate or ignorant or even (as Dawkins would have it) wicked?

As Christians have always stressed, life is a mystery. One might argue (in a toned-down version of Plantinga's argument against naturalism) that if we are the products of selection, then we should not expect to have adaptations that lead to knowledge of ultimate reality or nonreality. As Richard Dawkins of all people has so rightly said: "Modern physics teaches us that there is more to truth than meets the eye; or that meets the all too limited human mind, evolved as it was to cope with medium-sized objects moving at medium speeds through medium distances in Africa."[15] This does not necessarily mean that God is on the other side of the divide, but it does mean that Darwinians should not sneer at those who think that possibly he is. Perhaps there are things beyond—forever beyond—our ken. Perhaps therefore a little modesty or skepticism about our own nihilistic position is in order, along with a little more tolerance for those who might wish to make something more of the mystery of life.

Finally, creationists have a point when they argue that Darwinian science brings in cultural issues. Even though scientists may not necessarily be committed to the values of their culture, their science is impregnated with that culture. To say that metaphors can or ever will be eliminated is just nonsense. Doing so would cut off most of the important epistemic virtues of science. Every now and again, with good reason, evolutionists worry about the teleological nature of their theory—all of the talk about purpose and intention and function and so forth seems to imply a Mind Above that is guiding life below. Perhaps we can and should do away with such teleological language. But as Charles Darwin pointed out when people objected to his talk about the anthropomorphic notion of natural selection, metaphor and science are inseparable. "In the literal sense of the word, no doubt, natural selection is a false term; but who ever objected to chemists speaking of the elective affinities of the various elements?—and yet an acid cannot strictly be said to elect the base with which it in preference combines. It has been said that I speak of natural selection as an active power or Deity; but who objects to an author speaking of the attraction of gravity as ruling the movements of the planets?"[16]

The same is true today. Edward O. Wilson performed his miracles of understanding about the world of social insects precisely because he interpreted what he saw with the metaphor of a division of labor. Just looking at his ants would probably have led him nowhere. Because he examined them from an economic standpoint, asking what is the most sensible use of limited resources, he went a long way in unraveling the way that nature works. Likewise with arms races and selfish genes and everything else. The synthetic theory of evolution would have fizzled without Sewall Wright's brilliant metaphor of an adaptive landscape—and one can say this, even though today many people believe the metaphor of an adaptive landscape is not appropriate. Through use of the metaphor, science—good science—was done by Theodosius Dobzhansky and many others.

Although there is no necessary connection, given that evolutionists are members of society, one would be naive to think that metaphors will never give rise to value judgments. And it would be equally naive to think that such judgments will always be thought regrettable or inadvisable—that people will make them only by mistake or because they have been seduced into claiming something that they truly do not believe. The English philosopher Richard Braithwaite said that the price of the use of models is eternal vigilance, and the same is true of metaphors. But even if a scientist such as Geoffrey Parker does not exploit his metaphors to send a message, others surely will take them up and make more of them, deliberately, especially people in the public realm. One can write about evolutionary biology for general readers without sending a message, and often people do. But just as often others write about evolutionary biology precisely because they do have a message—a message about progress and the virtues of modern society—and they look to evolutionary biology as a way to express it.

Go to virtually any natural history museum, open any popular book on evolutionary theory—especially those written for children—and you will find the story of monad to man, amoebas to Americans. How often has one heard the story of life's tape, with everything squashed down into the period of a year? Around Christmas time the mammals appear, and late on New Year's Eve we get the great apes and Australopithecus, with us humans making the scene just as the bells are about to ring. Mention is never made of the late arrival of warthogs or poison ivy or the AIDS virus. Nor is there any false modesty about what has happened. We won, and we won because we are superior. (In a coda, readers are usually warned that if we do not take steps to control global warming and so forth we shall end with the Neanderthals and the dodo. Extinct.)

This use of evolutionary theory for the message it conveys so well is not necessarily wrong. Like Aldous Huxley, today's intellectuals are often somewhat wary of progress. It led to the bomb and to global warming and to overpopulation and much more that seems bad in our world. Scientists are among the minority of intellectuals who are almost universally optimistic. Even Gould, who denied biological progress, did so precisely because he believed in the possibility of social progress; indeed, denial of biological progress was his convoluted way of trying to advance social progress. Scientists buck the trend because (social constructivism notwithstanding) science is the one area of human experience that is unambiguously progressive. First Mendelian factors, next the classical theory of the gene, and then the double helix. Who would deny that, epistemologically, we have made progress? Unless you believe in scientific progress, you are going to flop as a scientist. You have got to push on and make advances, even when things seem darkest. Science stoppers are for theologians and philosophers, not for scientists.

In short, with science come thoughts of progress, and anyone who thinks that evolutionists—Gould not excepted—are unaffected by this is ignorant of evolutionists. Little wonder that the population geneticist John Maynard Smith joined with other scientists in being an enthusiast for the philosophy of the late Karl Popper. As we saw earlier, the one thing that Popper stressed above all else is that science yields objective, culture-value-free knowledge. There is a real world, independent of the scientist, and although we may never get to know this real world completely, in science we progress to ever-better understanding.[17] Evolutionists, by virtue of being scientists, tend toward progressivism, and they have in their hands a theory that lends itself precisely to such a philosophy. The slide from epistemological progress to the social and moral progress is quick and easy. Perhaps Edward O. Wilson and William D. Hamilton and others are atypical in thinking that their science has broader implications, but perhaps not.

Enough. I am a philosopher and historian, not an Old Testament prophet, and as such I do not criticize Wilson for promoting biodiversity or Hamilton for expressing his concerns about modern medical trends. (I suspect that Hamilton's cure might be worse than the disease, but, not being a social scientist, I am not in the business of offering ready solutions.) My area of expertise is the clash between evolutionists and creationists, and my analysis is that we have no simple clash between science and religion but rather between two religions. The outcome of the conflict is not obvious. Inasmuch as creationism is linked with premillennialism, and premillennialism is linked in America with the moral and political threats that many people perceive as impinging on our society, I do not see creationism vanishing any time soon. Given the horrendous events that have already happened in this new millennium, it would be naive to expect otherwise. But that is no excuse for inaction. Those of us who love science must do more than simply restate our positions or criticize the opposition. We must understand our own assumptions and, equally, find out why others have (often) legitimate concerns. This is not a plea for weak-kneed compromise but a more informed and self-aware approach to the issues. First understanding, and then some strategic moves. You now know why I wrote this book.

Notes

I. Christianity and Its Discontents

1. Livingstone 1988; Hastings, Mason, and Pyper 2000.
2. Bockmuehl 2001; Brown 1967; McGrath 1997.
3. McMullin 1986.
4. Lennox 2001.
5. Boyer 1992.
6. Tuveson 1949; Cohn 1961.
7. Herbert of Cherbury 1645, 86.
8. Kuhn 1957.
9. Kenny 2000.
10. Colley 1992.
11. Wesley, *Journal*, May 24, 1738.
12. Noll 2002; Holifield 2003.
13. Marty 1984.
14. Marsden 2003.
15. Firth 1979; Davidson 1977.
16. Bury 1920; Spadefora 1990.
17. Mommsen 1951.
18. Frankel 1948.
19. Condorcet 1795, 128.
20. Hartley 1764, 324–325.
21. Broughton 1768, 297. Millennialism is sometimes known as "millennarianism," although some restrict this term to premillennialism.
22. Tuveson 1968; Endy 1985.

23. Freneau 1902, I, 80–81.
24. Paine 1908, I, 268; Jefferson, May 1976. Jefferson's comment comes in a letter to Francis Adrian van de Kemp, July 30, 1816.

2. From Progress to Evolution

1. Ruse 1996.
2. Lovejoy 1936.
3. Diderot 1943, 48, 152.
4. Lamarck 1802, 38.
5. Darwin 1803, I, lines 309–314.
6. Darwin 1791, 3, lines 349–350.
7. Darwin 1794, 509.
8. Crosland 1967.
9. Darnton 1968.
10. Franklin et al. 1970, 124.
11. Coleman 1964; Outram 1984.
12. Cuvier 1817, I, 6.
13. Ibid., 67–68.
14. Ibid., 92.
15. Cuvier 1813.
16. Ruse 1999.

3. Growth of a Pseudoscience

1. Knox 1831, 486.
2. Richards 2003.
3. Rupke 1994.
4. Von Baer 1828, 37.
5. Owen 1843, 367–371.
6. Owen 1851, 448–450.
7. Owen 1849, 85–86; Darwin, *Species Notebook B*, 161, in Barrett et al. 1987.
8. Houghton 1957.
9. Secord 2000.
10. Shapin 1975.
11. Chambers 1844, 148–149, 360.
12. Chambers 1846, 400–402.
13. Powell 1855; Ross 1973.

14. Sedgwick 1845, 4; Brewster 1844, 503.
15. Turner 2002; Chadwick 1966.
16. Harrison 1979; Oliver 1978.
17. Miller 1856, 153.
18. Miller 1847, 105.
19. Turner 2002.
20. Newman 1973, 97.
21. Chadwick 1957.
22. Newman 1845, 64.
23. Newman 1868, V, xvii.
24. Whewell 1837, 3: 576, 588.
25. Clark and Hughes 1890, 2, 86.
26. Whewell 1853, 253, 243.

4. Charles Darwin

1. The best biography is Browne 1995, 2002.
2. Darwin 1842.
3. Darwin and Wallace 1858.
4. Darwin 1859, 63, 80–81.
5. Ibid., 446.
6. Ibid., 488, 490.
7. Ospovat 1981; Ruse 1996.
8. Darwin 1859, 129–130.
9. McMullin 1983.
10. Darwin 1959, 222.
11. Wallace 1870.
12. Letter to Asa Gray, May 22, 1860.

5. Failure of a Professional Science

1. Ellegård 1958.
2. Hull 1973.
3. Burchfield 1975.
4. Bates 1862; Wallace 1866.
5. Desmond 1994, 1997; Ruse 1996.

6. Hoppen 1998.
7. Harrison 1971.
8. Huxley 1900, I, 219.
9. Ibid., 2, 428.
10. Haeckel 1866.
11. Dupree 1959.
12. Lurie 1960.
13. Agassiz 1885, I, 369–370.
14. Huxley 1877, 90.
15. Lankester 1881.
16. Bowler 1996.
17. Rainger 1991; Winsor 1991.

6. Social Darwinism

1. Spencer 1904; Richards 1987.
2. Spencer 1857, 2–3.
3. Spencer 1851, 323–324.
4. Youmans 1879, 817.
5. Pusey 1983, 165, 77.
6. Ruse 2000; Bannister 1979.
7. Sumner 1914, 30.
8. Russett 1976, 94.
9. Pittenger 1993.
10. Letter from Engels to Marx, December 12, 1859.
11. Jones 1980, 73; Wallace 1905.
12. Kropotkin 1902, vi-viii, 293.
13. Bernhardi 1912, 10, quoted by Crook 1994, 83.
14. Friedlander 1997.
15. Gasman 1971, 168, 113n; Bullock 1991, 141.
16. Kellogg 1912, 140–141.
17. Russett 1989.
18. Darwin 1871, 2, 316, 326.
19. Allan 1869, cxcix; Ellis 1894, 246.
20. Allan 1869, cxcvii.
21. Cope 1887, 159.
22. Wallace 1900, 2, 507.
23. Lankester 1880, 60.

24. Wells 1895, 70.
25. Ibid., 101, 111, 101.
26. Kellogg 1912, 167–168.
27. Tyndall 1892, 2, 393.
28. Letter of April 4, 1882, quoted in Budd 1977, 92.
29. Quoted in Budd 1977, 181.
30. Oliver 1978; Harrison 1979.
31. Harrison 1969.
32. Owen 1858, I, xlii-iii.
33. Letter from T. H. Huxley to F. Dyster, January 30, 1859, Huxley Papers 15, 106.
34. White 2003.
35. December 1, 1890.

7. Christian Responses

1. Sedgwick 1860, 103.
2. Powell 1860, 139.
3. Duke of Argyll 1867, 293–294; Moore 1890.
4. Wright 1882, 219–220.
5. Temple 1884, 118–119.
6. McCosh 1887, I, 217.
7. McCosh 1871, 90–92.
8. Beecher 1885, 113.
9. McCosh 1890, 101.
10. Letter of March, 9, 1833, quoted in Hilton 1986, 244.
11. Altholz 1994.
12. Abbott and Campbell 1897, I, 291.
13. Murray 1996.
14. White 2003; Huxley 1870, 397; Arnold 1893, xi, xxxv, xxxvii.
15. Rauschenbush 1907, 6; Marsden 1990, 120.
16. Hammond 1918, 282.
17. Appleby 1999.
18. Letter of 1864, in Newman 1971, 266.
19. Newman 1873, 428–432, 389.
20. Elder 1996; letter of June 5, 1870, in Newman 1973, 137.
21. Mivart 1876, 198–199.
22. Pope John Paul II 1997.

8. Fundamentalism

1. Marty 1984; Noll 2002.
2. Marsden 2003; Holifield 2003.
3. Hickman 1865, 2, 557; I Corinthians 1:27.
4. Beecher 1861, 289.
5. Mark 10:21.
6. W. A. Stevens, quoted by Moorhead 1999, 36.
7. Ker 1988, 531.
8. Moody quoted in Marsden 1980, 35; James Henley Thornwell quoted in Noll 2002, 399. See also Numbers 1998; Roberts 1988.
9. Lurie 1960; Hodge 1872.
10. Grundlach 1995, 97; Hodge 1872, 1, 227; 2, 30.
11. H. F. Rall, quoted by Moorhead 1999, 44.
12. Goen 1959; Boyer 1992.
13. Numbers and Butler 1987; Stein 1999; Bull and Lockhart 1989.
14. Sandeen 1970.
15. Quoted in Weber 1983, 59.
16. Daniels 1877, 432; Smith 1971, 195.
17. Marsden 1980, 113.
18. Ibid., 98.
19. Numbers 1992a.
20. Dickson, Meyer, and Torrey [1910–1915], 1:4.
21. Kellogg 1917, 28–29.
22. Larson 1997.
23. Hunter 1914, 263.
24. Settle 1972, 108–109.
25. Numbers 1992b, 80.

9. Population Genetics

1. Bateson 1928, 390–391; Allen 1978; Rainger, Benson, and Maienschein 1988.
2. Provine 1971.
3. Box 1978.
4. Fisher 1930, 35.
5. Letter to Leonard Darwin, August 7, 1928, in Bennett 1983, 88.
6. Provine 1986.

7. Wright 1931, 158.
8. Wright 1932, 68.
9. Ibid., 168–169.
10. Fisher 1947, 1001.
11. Fisher 1950, 19–20.
12. Letter to J. T. McNeill, November 12, 1943.
13. Stapledon 1930, 319.
14. Henderson 1917, 138; interview with W. B. Provine, June 4, 1976; letters to Q. Wright, January 10, 1916, February 27, 1916.
15. Baker 1976.
16. Ford 1931.
17. Dobzhansky 1937, 134.
18. Ibid., 164.
19. Lewontin et al. 1981.
20. Letter to W. A. Gosline, March 5, 1948.
21. Letter to A. Cronquist, November 14, 1949.
22. Letter to G. G. Ferris, March 28, 1948.
23. Simpson 1949, 311, 315.
24. Joravsky 1970, 321.

10. Evolution Today

1. Freeman and Herron 1998.
2. Lewontin 1974.
3. Bowler 1996; Carroll, Grenier, and Weatherbee 2001; Raff 1996.
4. Gilbert, Opitz, and Raff 1996, 368.
5. Reeve and Sherman 1993, 18. See also Darwin 1862.
6. Hamilton 1964a, 1964b.
7. Dawkins 1976.
8. Hrdy 1999.
9. Eldredge and Gould 1972; Gould 2002.
10. Hull and Ruse 1998.
11. Dawkins 1997a.
12. Interview with author, May 1997.
13. Coyne 2002.
14. Dawkins 1997b, 399; Dawkins 1997a; Brockman 1995, 85–86.
15. Letter of May 22, 1860.
16. Gould 1999; Gould 1989, 318.

17. Wilson 1978, 188, 192.
18. Dawkins 2002.
19. Dawkins 2003, 160–161.
20. Dawkins 1997a.
21. Hamilton 2001, xlviii.
22. Interview with author, April 1988; Wilson 1992, 187.
23. Wilson 2002, 133.
24. Gould 1988, 1996.

11. Nature as Promise

1. Dawkins 1995, 133.
2. Wagar 1972, 225.
3. Webster 2000.
4. Teilhard de Chardin 1955, 101.
5. Ibid., 310–311.
6. Ibid., 312; "the human zoological group," 1956, quoted in Benz 1966, 216.
7. Medawar 1961; Gould 1980; Huxley 1959; Dobzhansky 1967.
8. Huxley 1912, 115–116.
9. Dawkins 1986, 189.
10. Dawkins 1997, 216–217.
11. Conway Morris 2003, 112.
12. Ibid., 196.
13. Conway Morris 2002, 161, 170.
14. Haught 2000, 76, 118, 117; his italics.
15. Ibid., 20–21, 118.
16. Haught 1993, 103; letter to author, March 2004.
17. Teilhard 1953, quoted in Benz 1966, 217–218.
18. Haught 2000, 155.
19. Haught 2004, 31; 1993, 106.
20. Ward 1996, 87.
21. Russell, Stoeger, and Coyne 1988.
22. Goodwin 2001.
23. Weber and Depew 2004, 180–181.
24. Ibid.
25. Rolston 1987, 111–112.
26. Ibid., 131.
27. Rolston 1999, 286, 288.

28. Letter to author, March 2004.
29. Rolston 2001, 59–60.

12. Earth's Last Days?

1. Numbers 1992.
2. Price 1902, 159.
3. Price 1923, 13; Numbers 2004, 84.
4. Numbers 1992, 85.
5. Price n.d., 1.
6. Webb 1994.
7. Tax 1960a, 1960b; Tax and Callender 1960; Smocovitis 1999.
8. Boyer 1992; Frykholm 2004.
9. Graham 1965, 165.
10. Whitcomb and Morris 1961, 175.
11. Morris 1983, 26–27.
12. Whitcomb and Morris 1961, 425.
13. Morris et al. 1976, iii; Morris 1989, 148.
14. DeHaan 1962, 101.
15. Ronald Reagan, speech in 1971, quoted in Boyer 1992, 142.
16. Ruse 1988.
17. Overton 1982.
18. Holden 2004.
19. Brooks 2001.
20. Frykholm 2004.
21. Forrest and Gross 2004; Behe 1996, 39.
22. Behe 1996, 70.
23. Dembski 1998a, 1998b.
24. Dembski 2002, 221.
25. Conversation with Larry Arnhart, spring 2003.
26. Shanks 2004; Dembski 1999, 206; Dembski 2001, 192.
27. Letter to author, June 2003; Johnson 1991, 157–158.
28. Johnson 1995, 37–38.
29. Plantinga 1991.
30. Letter to author, January 2005.
31. Nelson letter to author, January 2005; webpage of Avner and Rachel Boskey, www.davidstent.org/vision.htm; Meyer letter to author, December 2004.

32. Letters to author, June 2003.
33. Ibid.
34. Johnson 1997, 106.
35. Johnson 2000.

Conclusion

1. Livingstone 1996.
2. Hamilton 2001, xlviii.
3. Ibid., xi.
4. Johnson 1995, 150–151.
5. Morris 2002.
6. Camp 1999. The women's magazine is *Family Voice.*
7. Wilson 2002, 134.
8. Quoted by Johnson 1995, 189.
9. Quoted by Forrest and Gross 2004, 35.
10. Forrest and Gross 2004, 43–44.
11. Dembski 1999, 233.
12. Plantinga 1997, 152–153.
13. Miller 1999.
14. Budd 1977, 107.
15. Dawkins 2003, 19.
16. Darwin 1859.
17. Popper 1972, 1974.

References and Reading

1. Christianity and Its Discontents

Barnes, J., ed. 1984. *The Complete Works of Aristotle, Vol. I*. Princeton: Princeton University Press.

Bockmuehl, M., ed. 2001. *The Cambridge Companion to Jesus*. Cambridge: Cambridge University Press.

Boyer, P. 1992. *When Time Shall Be No More: Prophecy Belief in Modern American Culture*. Cambridge: Harvard University Press.

Broughton, T. 1768. *A Prospect of Futurity, in Four Dissertations on the Nature and Circumstances of the Life to Come*. London.

Brown, P. 1967. *Augustine of Hippo: A Biography*. London: Faber and Faber.

Bury, J. B. [1920] 1924. *The Idea of Progress: An Inquiry into Its Origin and Growth*. London: MacMillan.

Cohn, N. 1961. *The Pursuit of the Millennium: Revolutionary Messianism in Medieval and Reformation Europe, and Its Bearing on Modern Totalitarian Movements*. 2nd ed. New York: Harper.

Colley, L. 1992. *Britons: Forging the Nation, 1707–1837*. New Haven: Yale University Press.

Condorcet, A. N. [1795] 1956. *Sketch for a Historical Picture of the Progress of the Human Mind*. New York: Noonday Press.

Davidson, J. W. 1977. *The Logic of Millennial Thought: Eighteenth-Century New England*. New Haven: Yale University Press.

Endy, M. B. 1985. Just war, holy war, and millennialism in revolutionary America. *William and Mary Quarterly* 42: 3–25.

Firth, K. R. 1979. *The Apocalyptic Tradition in Reformation Britain, 1530–1645*. Oxford: Oxford University Press.

Frankel, C. [1948] 1969. *The Faith of Reason: The Idea of Progress in the French Enlighten-ment.* New York: Octagon.

Freneau, P. 1902. *Poems.* Princeton: Princeton University Press.

Hartley, T. 1764. *Paradise Restored: Or a Testament to the Doctrine of the Blessed Millennium.* London.

Hastings, A., A. Mason, and H. Pyper, eds. 2000. *The Oxford Companion to Christian Thought: Intellectual, Spiritual, and Moral Horizons of Christianity.* Oxford: Oxford University Press.

Herbert of Cherbury. 1645 [1944]. *De Religione Laici.* New Haven: Yale University Press.

Holifield, E. B. 2003. *Theology in America: Christian Thought from the Age of the Puritans to the Civil War.* New Haven: Yale University Press.

Kenny, A., ed. 2000. *Oxford History of Western Philosophy.* Oxford: Oxford University Press.

Kuhn, T. 1957. *The Copernican Revolution.* Cambridge: Harvard University Press.

Lennox, J. G. 2001. *Aristotle's Philosophy of Biology.* Cambridge: Cambridge University Press.

Livingstone, J. C. 1988. *Modern Christian Thought: In the Enlightenment and the Nineteenth Century.* Upper Saddle River, NJ: Prentice-Hall.

Marsden, G. M. 2003. *Jonathan Edwards: A Life.* New Haven: Yale University Press.

Marty, M. 1984. *Pilgrims in Their Own Land: 500 Years of Religion in America.* Boston: Little, Brown.

May, H. F. 1976. *The Enlightenment in America.* New York: Oxford University Press.

McGrath, A. E. 1997. *Christian Theology: An Introduction.* 2nd ed. Oxford: Blackwell.

McMullin, E. 1986. Introduction: evolution and creation. In *Evolution and Creation,* ed. E. McMullin, 1–58. Notre Dame: University of Notre Dame Press.

Mommsen, T. E. 1951. St. Augustine and the Christian idea of progress: the back-ground of the City of God. *Journal of the History of Ideas* 12: 346–74.

Noll, M. 2002. *America's God: From Jonathan Edwards to Abraham Lincoln.* New York: Oxford University Press.

Paine, T. 1908. *The Age of Reason.* Boston.

Paley, W. [1794] 1819. *Evidences of Christianity.* Collected Works, vol. 3. London: Rivington.

Paley, W. [1802] 1819. *Natural Theology.* Collected Works, vol. 4. London: Rivington.

Ruse, M. 2003. *Darwin and Design: Does Evolution Have a Purpose?* Cambridge: Harvard University Press.

Spadefora, D. 1990. *The Concept of Progress in Eighteenth Century Britain.* New Haven: Yale University Press.

Tuveson, E. L. 1949. *Millennium and Utopia: A Study in the Background of the Idea of Progress.* Berkeley: University of California Press.

———. 1968. *Redeemer Nation: The Idea of America's Millennial Role.* Chicago: University of Chicago Press.

2. From Progress to Evolution

Burkhardt, R. W. 1995. *The Spirit of System: Lamarck and Evolutionary Biology.* Cambridge: Harvard University Press.

Coleman, W. 1964. *Georges Cuvier Zoologist: A Study in the History of Evolution Theory.* Cambridge: Harvard University Press.

Crosland, M. 1967. *The Society of Arcueil: A View of French Science at the Time of Napoleon I.* London: Heinemann.

Cuvier, G. [1813] 1822. *Theory of the Earth.* 4th ed., ed. Robert Jameson. Edinburgh: William Blackwood.

———. 1817. *Le règne animal distribué d'aprés son organisation, pour servir de base à l'histoire naturelle des animaux et d'introduction à l'anatomie comparée.* Paris.

Darnton, R. 1968. *Mesmerism and the End of the Enlightenment in France.* Cambridge: Harvard University Press.

Darwin, E. 1791. *The Botanic Garden: Part I, The Economy of Vegetation.* London: J. Johnson.

———. 1794. *Zoonomia; or, The Laws of Organic Life.* London: J. Johnson.

———. 1803. *The Temple of Nature.* London: J. Johnson.

Diderot, D. 1796 [1972]. *The Nun.* London: Penguin.

———. 1943. *Diderot: Interpreter of Nature.* New York: International Publishers.

Franklin, B., et al. 1970. Report of the commissioners charged by the King to examine animal magnetism. In *Foundations of Hypnosis from Mesmer to Freud,* ed. M. N. Pinterow, 89–124. Springfield, IL: Charles C. Thomas.

Lamarck, J. B. 1802. *Recherches sur L'organisation des corps vivans.* Paris: Maillard.

———. 1809. *Philosophie zoologique.* Paris: Dentu.

———. 1820. *Systeme analytique des connaissances positives de l'homme, restreintes a celles qui proviennent directement ou indirectement de l'observation.* Paris: Belin.

Lovejoy, A. O. 1936. *The Great Chain of Being.* Cambridge: Harvard University Press.

Outram, D. 1984. *Georges Cuvier: Vocation, Science and Authority in Post Revolutionary France.* Manchester: Manchester University Press.

Ruse, M. 1996. *Monad to Man: The Concept of Progress in Evolutionary Biology.* Cambridge: Harvard University Press.

————. 1999. *The Darwinian Revolution: Science Red in Tooth and Claw.* 2nd ed. Chicago: University of Chicago Press.

3. Growth of a Pseudoscience

Barrett, P. H., P. J. Gautrey, S. Herbert, D. Kohn, and S. Smith, eds. 1987. *Charles Darwin's Notebooks, 1836–1844.* Ithaca: Cornell University Press.

Brewster, D. 1844. Vestiges. *North British Review* 3: 470–515.

————. 1854. *More Worlds than One: The Creed of the Philosopher and the Hope of the Christian.* London: Camden Hotten.

Chadwick, O. 1957. *From Bossuet to Newman: The Idea of Doctrinal Development.* Cambridge: Cambridge University Press.

————. 1966. *The Victorian Church. Part I.* London: A. and C. Black.

Chambers, R. 1844. *Vestiges of the Natural History of Creation.* London: Churchill.

————. 1846. *Vestiges of the Natural History of Creation.* 5th ed. London: J. Churchill.

Clark, J. W., and T. M. Hughes, eds. 1890. *Life and Letters of the Reverend Adam Sedgwick.* Cambridge: Cambridge University Press.

Harrison, J. F. C. 1979. *The Second Coming: Popular Millenarianism, 1780–1850.* New Brunswick: Rutgers University Press.

Houghton, W. E. 1957. *The Victorian Frame of Mind.* New Haven: Yale University Press.

Knox, R. 1831. Observations on the structure of the stomach of the Peruvian llama; to which are prefixed remarks on the analogical reasoning of anatomists, in the determination *a priori* of unknown species and unknown structures. *Transactions of the Royal Society of Edinburgh* 11: 479–498.

Lyell, C. 1830–1833. *Principles of Geology: Being an Attempt to Explain the Former Changes in the Earth's Surface by Reference to Causes Now in Operation.* London: John Murray.

Miller, H. 1847. *Footprints of the Creator; Or, the Asterolepis of Stromness.* Edinburgh: Constable.

————. 1856. *The Testimony of the Rocks, or Geology in its Bearings on the Two Theologies, Natural and Revealed.* Edinburgh: Constable.

Newman, J. H. [1845] 1989. *An Essay on the Development of Christian Doctrine.* Notre Dame: Notre Dame University Press.

————. 1868. *Parochial and Plain Sermons.* London: Rivington.

————. 1973. *The Letters and Diaries of John Henry Newman,* vol. 25, ed. C. S. Dessain and T. Gornall. Oxford: Clarendon Press.

Oliver, W. H. 1978. *Prophets and Millennialists: The Uses of Biblical Prophecy in England from the 1790s to the 1840s.* Auckland: Auckland University Press.

Owen, R. [1837] 1992. *The Hunterian Lectures in Comparative Anatomy, May and June 1837,* ed. P. R. Sloan. Chicago: Chicago University Press.

———. 1843. *Lectures on the Comparative Anatomy and Physiology of the Invertebrate Animals.* London: Longman, Brown, Green and Longmans.

———. 1849. *On the Nature of Limbs.* London: Voorst.

———. 1851. Principles of geology by Sir Charles Lyell, etc. *Quarterly Review* 89: 412–451.

Powell, B. 1855. *Essays on the Spirit of the Inductive Philosophy.* London: Longman, Brown, Green, and Longmans.

Richards, R. J. 2003. *The Romantic Conception of Life: Science and Philosophy in the Age of Goethe.* Chicago: University of Chicago Press.

Ross, R. H. 1973. *Alfred, Lord Tennyson, "In Memoriam": An Authoritative Text, Backgrounds and Sources of Criticism.* New York: Norton.

Rupke, N. A. 1994. *Richard Owen: Victorian Naturalist.* New Haven: Yale University Press.

Ruse, M. 1999. *The Darwinian Revolution: Science Red in Tooth and Claw.* 2nd ed. Chicago: University of Chicago Press.

Secord, J. A. 2000. *Victorian Sensation: The Extraordinary Publication, Reception, and Secret Authorship of Vestiges of the Natural History of Creation.* Chicago: University of Chicago Press.

Sedgwick, A. 1845. Vestiges. *Edinburgh Review* 82: 1–85.

Shapin, S. 1975. Phrenological knowledge and the social structure of early nineteenth century Edinburgh. *Annals of Science* 32: 219–243.

Turner, F. M. 2002. *John Henry Newman: The Challenge to Evangelical Religion.* New Haven: Yale University Press.

von Baer, K. E. [1828–1837] 1853. Über Enwickelungsgeschichte der Thiere: fragments related to *Philosophical Zoology* selected from the works of K. E. von Baer. In *Scientific Memoirs,* ed. and trans. A. Henfry and T. H. Huxley. London: Taylor and Francis.

Whewell, W. 1837. *The History of the Inductive Sciences.* London: Parker.

———. 1840. *The Philosophy of the Inductive Sciences.* London: Parker.

———. 1845. *Indications of the Creator.* London: Parker.

———. 1853 [2001]. *Of the Plurality of Worlds: A Facsimile of the First Edition of 1853, Plus Previously Unpublished Material Excised by the Author Just Before the Book Went to Press; and Whewell's Dialogue Rebutting His Critics, Reprinted from the Second Edition.* Chicago: University of Chicago Press.

4. Charles Darwin

Browne, J. 1995. *Charles Darwin: Voyaging.* New York: Knopf.

———. 2002. *Charles Darwin: The Power of Place.* New York: Knopf.

Darwin, C. 1839. *Journal of Researches into the Geology and Natural History of the Various Countries Visited by HMS Beagle.* London: Henry Colburn.

———. 1842. *The Structure and Distribution of Coral Reefs.* London: Smith Elder.

———. 1859. *On the Origin of Species.* London: John Murray.

———. 1862. *On the Various Contrivances by Which British and Foreign Orchids Are Fertilized by Insects, and On the Good Effects of Intercrossing.* London: John Murray.

———. 1871. *The Descent of Man.* London: John Murray.

———. 1959. *The Origin of Species by Charles Darwin: A Variorum Text,* ed. M. Peckham. Philadelphia: University of Pennsylvania Press.

Darwin, C., and A. R. Wallace. 1858. On the tendency of species to form varieties; and on the perpetuation of varieties and species by means of selection. *Proceedings of the Linnaean Society, Zoological Journal* 3: 46–62.

Lyell, C. 1830–1833. *Principles of Geology: Being an Attempt to Explain the Former Changes in the Earth's Surface by Reference to Causes Now in Operation.* London: John Murray.

Malthus, T. R. [1826] 1914. *An Essay on the Principle of Population.* 6th ed. London: Everyman.

McMullin, E. 1983. Values in science. *PSA 1982,* ed. P. D. Asquith, and T. Nickles, 3–28. East Lansing: Philosophy of Science Association.

Ospovat, D. 1981. *The Development of Darwin's Theory: Natural History, Natural Theology, and Natural Selection, 1838–1859.* Cambridge: Cambridge University Press.

Paley, W. [1794] 1819. *Evidences of Christianity.* Collected Works, vol. 3. London: Rivington.

———. [1802] 1819. *Natural Theology.* Collected Works, vol. 4. London: Rivington.

Ruse, M. 1975. Darwin's debt to philosophy: an examination of the influence of the philosophical ideas of John F. W. Herschel and William Whewell on the development of Charles Darwin's theory of evolution. *Studies in History and Philosophy of Science* 6: 159–181.

———. 1996. *Monad to Man: The Concept of Progress in Evolutionary Biology.* Cambridge: Harvard University Press.

Wallace, A. R. 1870. *Contributions to the Theory of Natural Selection: A Series of Essays.* London: Macmillan.

Whewell, W. 1837. *The History of the Inductive Sciences.* London: Parker.

5. Failure of a Professional Science

Agassiz, E. C., ed. 1885. *Louis Agassiz: His Life and Correspondence.* Boston: Houghton Mifflin.

Bates, H. W. 1862. Contributions to an insect fauna of the Amazon Valley Lepidoptera: Heliconidae. *Transactions of the Linnaean Society of London* 23:495–515.

Bowler, P. 1996. *Life's Splendid Drama.* Chicago: University of Chicago Press.

Burchfield, J. D. 1975. *Lord Kelvin and the Age of the Earth.* New York: Science History Publications.

Desmond, A. 1994. *Huxley, the Devil's Disciple.* London: Michael Joseph.

———. 1997. *Huxley, Evolution's High Priest.* London: Michael Joseph.

Dupree, A. H. 1959. *Asa Gray, 1810–1888.* Cambridge: Harvard University Press.

Ellegård, A. 1958. *Darwin and the General Reader.* Goteborg: Goteborgs Universitets Arsskrift.

Haeckel, E. 1866. *Generelle Morphologie der Organismen.* Berlin: Georg Reimer.

Harrison, J. F. C. 1971. *The Early Victorians, 1823–51.* London: Weidenfeld and Nicolson.

Hoppen, K. T. 1998. *The Mid-Victorian Generation, 1846–1886.* Oxford: Oxford University Press.

Hull, D., ed. 1973. *Darwin and His Critics.* Cambridge: Harvard University Press.

Huxley, L. 1900. *The Life and Letters of Thomas Henry Huxley.* London: Macmillan.

Huxley, T. H. 1854. Vestiges, etc. *British and Foreign Medico-Chirurgical Review* 13: 425–439.

———. 1863. *Evidence as to Man's Place in Nature.* London: Williams and Norgate.

———. 1877. *American Addresses, with a Lecture on the Study of Biology.* London: Macmillan.

Lankester, E. R. 1881. Limulus an Arachnid. *Quarterly Journal of Microscopical Science* n.s. 21: 609–649.

Lurie, E. 1960. *Louis Agassiz: A Life in Science.* Chicago: University of Chicago Press.

Rainger, R. 1991. *An Agenda for Antiquity: Henry Fairfield Osborn and Vertebrate Paleontology at the American Museum of Natural History, 1890–1935.* Tuscaloosa: University of Alabama Press.

Ruse, M. 1996. *Monad to Man: The Concept of Progress in Evolutionary Biology.* Cambridge: Harvard University Press.

———. 1999. *The Darwinian Revolution: Science Red in Tooth and Claw.* 2nd ed. Chicago: University of Chicago Press.

Wallace, A. R. [1866] 1870. On the phenomena of variation and geographical distribution as illustrated by the Papilionidae of the Malayan region. *Transactions of the Linnaean Society of London* 25: 1–27.

Winsor, M. P. 1991. *Reading the Shape of Nature: Comparative Zoology at the Agassiz Museum.* Chicago: University of Chicago Press.

6. Social Darwinism

Allan, J. M. 1869. On the real differences in the minds of men and women. *Journal of Anthropology* 7:cxcv–ccxix.

Bannister, R. 1979. *Social Darwinism: Science and Myth in Anglo-American Social Thought.* Philadelphia: Temple University Press.

Budd, S. 1977. *Varieties of Unbelief: Atheists and Agnostics in English Society 1850–1960.* London: Heinemann.

Bullock, A. 1991. *Hitler and Stalin: Parallel Lives.* London: HarperCollins.

Cope, E. D. 1887. *The Origin of the Fittest: Essays in Evolution.* New York: Macmillan.

Crook, P. 1994. *Darwinism: War and History.* Cambridge: University of Cambridge Press.

Darwin, C. 1871. *The Descent of Man.* London: John Murray.

Ellis, H. 1894. *Man and Woman: A Study of Human Secondary Sex Characteristics.* London: Walter Scott.

Farrar, F. W. 1858. *Eric or, Little by Little.* London: Black.

Friedlander, S. 1997. *Nazi Germany and the Jews: The Years of Persecution, 1933–39.* London: Weidenfeld and Nicolson.

Gasman, D. 1971. *The Scientific Origins of National Socialism: Social Darwinism in Ernst Haeckel and the Monist League.* New York: Elsevier.

Harrison, J. F. C. 1969. *Quest for the New Moral World: Robert Owen and the Owenites in Britain and America.* New York: Scribner's.

———. 1979. *The Second Coming: Popular Millenarianism 1780–1850.* New Brunswick: Rutgers University Press.

Huxley, J. S. 1927. *Religion Without Revelation.* London: Ernest Benn.

Huxley, T. H. 1891. *Social Diseases and Worse Remedies.* London: Macmillan.

———. 1893. Evolution and ethics. In *Evolution and Ethics,* 46–116. London: Macmillan.

Jones, G. 1980. *Social Darwinism and English Thought.* Brighton: Harvester.

Kellogg, V. L. 1912. *Beyond War: A Chapter in the Natural History of Man.* New York: Henry Holt.

Kropotkin, P. [1902] 1955. *Mutual Aid.* Boston: Extending Horizons Books.

Lankester, E. R. 1880. *Degeneration: A Chapter in Darwinism.* London: Macmillan.

Moore, G. E. 1903. *Principia Ethica.* Cambridge: Cambridge University Press.

Oliver, W. H. 1978. *Prophets and Millennialists: The Uses of Biblical Prophecy in England from the 1790s to the 1840s.* Auckland: Auckland University Press.

Owen, R. 1858. *A Supplementary Appendix to the First Volume of the Life of Robert Owen. Containing a Series of Reports, Addresses, Memorials, and Other Documents Referred to in that Volume, 1808–1820.* London.

Pittenger, M. 1993. *American Socialists and Evolutionary Thought, 1870–1920.* Madison: University of Wisconsin Press.

Pusey, J. R. 1983. *China and Charles Darwin.* Cambridge: Harvard University Press.

Richards, R. J. 1987. *Darwin and the Emergence of Evolutionary Theories of Mind and Behavior.* Chicago: University of Chicago Press.

Ruse, M. 1996. *Monad to Man: The Concept of Progress in Evolutionary Biology.* Cambridge: Harvard University Press.

———. 2000. *The Evolution Wars: A Guide to the Controversies.* Santa Barbara: ABC-CLIO.

Russett, C. E. 1976. *Darwin in America: The Intellectual Response, 1865–1912.* San Francisco: Freeman.

———. 1989. *Sexual Science: The Victorian Construction of Womanhood.* Cambridge: Harvard University Press.

Spencer, H. 1851. *Social Statics; Or the Conditions Essential to Human Happiness Specified and the First of Them Developed.* London: J. Chapman.

———. 1857. Progress: its law and cause. *Westminster Review* 67: 244–267.

———. 1862. *First Principles.* London: Williams and Norgate.

———. 1904. *Autobiography.* London: Williams and Norgate.

Sumner, W. G. 1914. *The Challenge of Facts and Other Essays.* New Haven: Yale University Press.

Tyndall, J. 1892. *Fragments of Science.* London: Longmans, Green.

von Berhardi, F. 1912. *Germany and the Next War.* London: Edward Arnold.

Wallace, A. R. 1900. *Studies: Scientific and Social.* London: Macmillan.

———. 1905. *My Life: A Record of Events and Opinions.* London: Chapman and Hall.

Wells, H. G. 1895. *The Time Machine.* London: Heinemann.

White, A. D. 1896. *History of the Warfare of Science with Theology in Christendom.* New York: Appleton.

White, P. 2003. *Thomas Huxley: Making the Man of Science.* Cambridge: University of Cambridge Press.

Youmans, W. J. 1879. Science in relation to war. *Popular Science* 14: 817–819.

7. Christian Responses

Abbott, E., and L. Campbell, eds. 1897. *The Life and Letters of Benjamin Jowett.* London: John Murray.

Altholz, J. L. 1994. *Anatomy of a Controversy: The Debate over "Essays and Reviews."* Aldershot: Scolar Press.

Appleby, R. S. 1999. Exposing Darwin's "hidden agenda": Roman Catholic responses to evolution, 1875–1925. In *Disseminating Darwinism: The Role of Place, Race, Religion, and Gender,* ed. R. L. Numbers and J. Stenhouse, 173–208. Cambridge: Cambridge University Press.

Argyll, Duke of. 1867. *The Reign of Law.* London: Alexander Strahan.

Arnold, M. 1893. *God and the Bible.* New York: Macmillan.

Beecher, H. W. 1885. *Evolution and Religion.* New York: Fords, Howard, and Hulbert.

Elder, G. P. 1996. *Chronic Vigour: Darwin, Anglicans, Catholics, and the Development of a Doctrine of Providential Evolution.* Lanham, MD: University Press of America.

Gray, A. 1876. *Darwiniana.* New York: Appleton.

Hammond, W. E. 1918. The end of the world. *Biblical World* 51: 282.

Hilton, B. 1986. *The Age of Atonement: The Influence of Evangelicalism on Social and Economic Thought, 1785-1865.* Oxford: Oxford University Press.

Huxley, T. H. 1870. The school boards: what they can do, and what they may do. *Science and Education. Collected Essays,* vol. 3. London: Macmillan.

John Paul II. 1997. The Pope's message on evolution. *Quarterly Review of Biology* 72: 377–383.

Marsden, G. M. 1980. *Fundamentalism and American Culture: The Shaping of Twentieth Century Evangelicalism, 1870–1925.* Oxford: Oxford University Press.

———. 1990. *Religion and American Culture.* San Diego: Harcourt, Brace, and Jovanovich.

McCosh, J. 1871. *Christianity and Positivism: A Series of Lectures to the Times on Natural Theology and Christian Apologetics.* London: Macmillan.

———. 1887. *Realistic Philosophy Defended in a Philosophic Series.* New York.

———. 1890. *The Religious Aspect of Evolution.* New York: Scribner's.

Mivart, St. G. 1871. *Genesis of Species.* London: Macmillan.

———. 1876. *Contemporary Evolution.* London: Henry S. King.

Moore, A. 1890. The Christian doctrine of God. *Lux Mundi,* ed. C. Gore. London: John Murray.

Murray, N. 1996. *A Life of Matthew Arnold.* New York: St. Martin's Press.

Newman, J. H. [1873] 1999. *The Idea of a University.* New York: Regnery.

———. 1971. *The Letters and Diaries of John Henry Newman,* vol. 21, ed. C. S. Dessain and T. Gornall. Edinburgh: Thomas Nelson.

———. 1973. *The Letters and Diaries of John Henry Newman,* vol. 25, ed. C. S. Dessain and T. Gornall. Oxford: Clarendon Press.

Powell, B. 1860. On the study of the evidences of Christianity. *Essays and Reviews*, 94–144. London: Longman, Green, Longman, and Roberts.

Rauschenbush, W. 1907 [1964]. *Christianity and the Social Crisis.* New York: Harper and Row.

Sedgwick, A. [1860] 1988. Objections to Mr. Darwin's theory of the origin of species in *The Spectator*, April 7, 1860. In *But Is It Science? The Philosophical Question in the Creation/Evolution Controversy*, ed. M. Ruse, 99–105. Buffalo: Prometheus.

Temple, F. 1884. *The Relations Between Religion and Science.* London: Macmillan and Co.

White, P. 2003. *Thomas Huxley: Making the Man of Science.* Cambridge: University of Cambridge Press.

Wright, G. F. 1882. *Studies in Science and Religion.* Andover, MA: Draper.

8. Fundamentalism

Beecher, H. W. 1861. *Fast Day Sermons: Or, The Pulpit on the State of the Country.* New York.

Boyer, P. 1992. *When Time Shall Be No More: Prophecy Belief in Modern American Culture.* Cambridge: Harvard University Press.

Bryan, W. J. 1922. *In His Image.* New York: Fleming H. Revell.

Bull, M., and K. Lockhart. 1989. *Seeking a Sanctuary: Seventh-day Adventism and the American Dream.* New York: Harper and Row.

Daniels, W. H. 1877. *Moody: His Words, Works, and Workers.* New York: Nelson and Phillips.

Dickson, A. C., L. Meyer, and R. A. Torrey, eds. [1910–1915.] *The Fundamentals: A Testimony.* 12 vols. Chicago: Testimony Publishing Co.

Goen, C. C. 1959. Jonathan Edwards: a new departure in eschatology. *Church History* 28: 25–40.

Grundlach, B. J. 1995. "The evolution question at Princeton, 1845–1929." Ph.D. diss., University of Rochester.

Hickman, E., ed. 1865. *The Works of Jonathan Edwards.* London.

Hodge, C. 1872. *Systematic Theology.* London and Edinburgh: Nelson.

———. 1874. *What Is Darwinism?* New York: Scribner's.

Holifield, E. B. 2003. *Theology in America: Christian Thought from the Age of the Puritans to the Civil War.* New Haven: Yale University Press.

Hunter, G. 1914. *A Civic Biology: Presented in Problems.* New York: American Book Company.

Kellogg, V. L. 1917. *Headquarters Nights: A Record of Conversations and Experiences at the Headquarters of the German Army in France and Belgium.* Boston: Atlantic Monthly Press.

Ker, I. 1988. *John Henry Newman: A Biography.* Oxford: Oxford University Press.

Larson, E. J. 1997. *Summer for the Gods: The Scopes Trial and America's Continuing Debate over Science and Religion.* New York: Basic Books.

Lurie, E. 1960. *Louis Agassiz: A Life in Science.* Chicago: University of Chicago Press.

Marsden, G. M. 1980. *Fundamentalism and American Culture: The Shaping of Twentieth Century Evangelicalism, 1870–1925.* Oxford: Oxford University Press.

———. 2003. *Jonathan Edwards: A Life.* New Haven: Yale University Press.

Marty, M. 1984. *Pilgrims in Their Own Land: 500 Years of Religion in America.* Boston: Little, Brown.

Moorhead, J. H. 1999. *World without End: Mainstream American Protestant Visions of the Last Things, 1880–1925.* Bloomington: Indiana University Press.

Noll, M. 2002. *America's God: From Jonathan Edwards to Abraham Lincoln.* New York: Oxford University Press.

Numbers, R. L. 1992a. *Prophetess of Health: Ellen G. White and the Origins of Seventh-day Adventist Health Reform.* Rev. ed. Knoxville: University of Tennessee Press.

———. 1992b. *The Creationists: The Evolution of Scientific Creationism.* New York: Knopf.

———. 1998. *Darwinism Comes to America.* Cambridge: Harvard University Press.

Numbers, R. L., and J. M. Butler, eds. 1987. *The Disappointed: Millerism and Millenarianism in the Nineteenth Century.* Bloomington: Indiana University Press.

Roberts, J. H. 1988. *Darwinism and the Divine in America: Protestant Intellectuals and Organic Evolution, 1859–1900.* Madison: University of Wisconsin Press.

Sandeen, E. R. 1970. *The Roots of Fundamentalism: British and American Millenarianism.* Chicago: University of Chicago Press.

Settle, M. L. 1972. *The Scopes Trial: The State of Tennessee v. John Thomas Scopes.* New York: Franklin Watts.

Smith, W. M. 1971. *The Best of D. L. Moody.* Chicago: Moody Press.

Stein, S. J. 1999. Apocalypticism outside the mainstream. *The Encyclopedia of Apocalypticism:* Vol. 3, *Apocalypticism in the Modern Period and Contemporary Age,* ed. S. J. Stein, 108–139. New York: Continuum.

Weber, T. P. 1983. *Living in the Shadow of the Second Coming: American Premillennialism, 1875–1982.* Chicago: University of Chicago Press.

9. Population Genetics

Allen, G. E. 1978. *Life Science in the Twentieth Century.* Cambridge: Cambridge University Press.

Baker, J. R. 1976. Julian Sorell Huxley. *Biographical Memoirs of Fellows of the Royal Society* 22: 207–238.

Bateson, B. 1928. *William Bateson, F. R. S., Naturalist: His Essays and Addresses together with a Short Account of His Life.* Cambridge: Cambridge University Press.

Bennett, J. H., ed. 1983. *Natural Selection, Heredity, and Eugenics. Including Selected Correspondence of R. A. Fisher with Leonard Darwin and Others.* Oxford: Oxford University Press.

Box, J. F. 1978. *R. A. Fisher: The Life of a Scientist.* New York: Wiley.

Dobzhansky, T. 1937. *Genetics and the Origin of Species.* New York: Columbia University Press.

Fisher, R. A. 1930. *The Genetical Theory of Natural Selection.* Oxford: Oxford University Press.

———. 1947. The renaissance of Darwinism. *Listener* 37: 1001.

———. 1950. *Creative Aspects of Natural Law.* The Eddington Memorial Lecture. Cambridge: Cambridge University Press.

Ford, E. B. 1931. *Mendelism and Evolution.* London: Methuen.

Henderson, L. J. 1913. *The Fitness of the Environment.* New York: Macmillan.

———. 1917. *The Order of Nature.* Cambridge: Harvard University Press.

Huxley, J. S. 1942. *Evolution: The Modern Synthesis.* London: Allen and Unwin.

Joravsky, D. 1970. *The Lysenko Affair.* Cambridge: Harvard University Press.

Lewontin, R. C., J. A. Moore, W. B. Provine, and B. Wallace, eds. 1981. *Dobzhansky's Genetics of Natural Populations I-XLIII.* New York: Columbia University Press.

Provine, W. B. 1971. *The Origins of Theoretical Population Genetics.* Chicago: University of Chicago Press.

———. 1986. *Sewall Wright and Evolutionary Biology.* Chicago: University of Chicago Press.

Rainger, R., K. R. Benson, and J. Maienschein, eds. 1988. *The American Development of Biology.* Philadelphia: University of Pennsylvania Press.

Ruse, M. 1996. *Monad to Man: The Concept of Progress in Evolutionary Biology.* Cambridge: Harvard University Press.

———. 1999. *Mystery of Mysteries: Is Evolution a Social Construction?* Cambridge: Harvard University Press.

Simpson, G. G. 1944. *Tempo and Mode in Evolution.* New York: Columbia University Press.

———. 1949. *The Meaning of Evolution.* New Haven: Yale University Press.

———. 1953. *The Major Features of Evolution.* New York: Columbia University Press.

Stapledon, W. O. 1930. *Last and First Men: A Story of the Near and Far Future.* London: Methuen.

Wright, S. 1931. Evolution in Mendelian populations. *Genetics* 16: 97–159.

———. 1932. The roles of mutation, inbreeding, crossbreeding and selection in evolution. *Proceedings of the Sixth International Congress of Genetics* I: 356–366.

10. Evolution Today

Bowler, P. 1996. *Life's Splendid Drama.* Chicago: University of Chicago Press.

Brockman, J. 1995. *The Third Culture: Beyond the Scientific Revolution.* New York: Simon and Schuster.

Carroll, S. B., J. K. Grenier, and S. D. Weatherbee. 2001. *From DNA to Diversity: Molecular Genetics and the Evolution of Animal Design.* Oxford: Blackwell.

Coyne, J. 2002. Intergalactic Jesus. *London Review of Books,* May 9.

Darwin, C. 1862. *On the Various Contrivances by Which British and Foreign Orchids Are Fertilized by Insects, and On the Good Effects of Intercrossing.* London: John Murray.

Dawkins, R. 1976. *The Selfish Gene.* Oxford: Oxford University Press.

———. 1997a. Is science a religion? *The Humanist* 57, no. 1.

———. 1997b. Obscurantism to the rescue. *Quarterly Review of Biology* 72: 397–399.

———. 1997c. Religion is a virus. *Mother Jones,* November 1.

———. 2002. An open letter to Prince Charles. In *Genetically Modified Foods,* ed. M. Ruse and D. Castle, 16–19. Buffalo: Prometheus.

———. 2003. *A Devil's Chaplain: Reflections on Hope, Lies, Science and Love.* Boston: Houghton Mifflin.

Eldredge, N., and S. J. Gould. 1972. Punctuated equilibria: an alternative to phyletic gradualism. In *Models in Paleobiology,* ed. T. J. M. Schopf, 82–115. San Francisco: Freeman.

Freeman, S., and J. C. Herron. 1998. *Evolutionary Analysis.* Englewood Cliffs: Prentice-Hall.

Gilbert, S. F., J. M. Opitz, and R. A. Raff. 1996. Resynthesizing evolutionary and developmental biology. *Developmental Biology* 173: 357–372.

Gould, S. J. 1988. On replacing the idea of progress with an operational notion of directionality. In *Evolutionary Progress,* ed. M. H. Nitecki, 319–338. Chicago: University of Chicago Press.

———. 1989. *Wonderful Life: The Burgess Shale and the Nature of History.* New York: Norton.

———. 1996. *Full House: The Spread of Excellence from Plato to Darwin.* New York: Paragon.

————. 1999. *Rocks of Ages: Science and Religion in the Fullness of Life.* New York: Ballantine.

————. 2002. *The Structure of Evolutionary Theory.* Cambridge: Harvard University Press.

Hamilton, W. D. 1964a. The genetical evolution of social behaviour I. *Journal of Theoretical Biology* 7: 1–16.

————. 1964b. The genetical evolution of social behaviour II. *Journal of Theoretical Biology* 7: 17–32.

————. 2001. *Narrow Roads of Gene Land: The Collected Papers of W. D. Hamilton.* Vol. 2, *Evolution of Sex.* Oxford: Oxford University Press.

Hrdy, S. 1999. *Mother Nature.* New York: Pantheon.

Hull, D. L., and M. Ruse, eds. 1998. *Readings in the Philosophy of Biology: Oxford Readings in Philosophy.* Oxford: Oxford University Press.

Lewontin, R. C. 1974. *The Genetic Basis of Evolutionary Change.* New York: Columbia University Press.

Raff, R. 1996. *The Shape of Life: Genes, Development, and the Evolution of Animal Form.* Chicago: University of Chicago Press.

Reeve, H. K., and P. W. Sherman. 1993. Adaptation and the goals of evolutionary research. *Quarterly Review of Biology* 68: 1–32.

Ruse, M. 1999. *Mystery of Mysteries: Is Evolution a Social Construction?* Cambridge: Harvard University Press.

Wilson, E. O. 1975. *Sociobiology: The New Synthesis.* Cambridge: Harvard University Press.

————. 1978. *On Human Nature.* Cambridge: Harvard University Press.

————. 1992. *The Diversity of Life.* Cambridge: Harvard University Press.

————. 2002. *The Future of Life.* New York: Vintage Books.

11. Nature as Promise

Benz, E. 1966. *Evolution and Christian Hope: Man's Concept of the Future, from the Early Fathers to Teilhard de Chardin.* Garden City: Doubleday.

Conway Morris, Simon. 2002. Does biology have an eschatology, and if so does it have cosmological implications? In *The Far-Future Universe: Eschatology from a Cosmic Perspective* ed. G. Ellis, 158–174. Philadelphia: Templeton Foundation Press.

————. 2003. *Life's Solution: Inevitable Humans in a Lonely Universe.* Cambridge: Cambridge University Press.

Dawkins, R. 1986. *The Blind Watchmaker.* New York: Norton.

————. 1995. *River Out of Eden: A Darwinian View of Life.* New York: Basic Books.

————. 1997. Human chauvinism: Review of *Full House* by Stephen Jay Gould. *Evolution* 51, no. 3: 1015–20.

Dobzhansky, T. 1967. *The Biology of Ultimate Concern.* New York, N.Y.: The New American Library, Inc.

Goodwin, B. 2001. *How the Leopard Changed Its Spots,* 2nd ed. Princeton: Princeton University Press.

Gould, S. J. 1980. The Piltdown conspiracy. *Natural History* 89, 8–28.

————. 1989. *Wonderful Life: The Burgess Shale and the Nature of History.* New York, N.Y.: W. W. Norton Co.

Haught, J. F. 1993. *The Promise of Nature.* New York: Paulist Press.

————. 2000. *God after Darwin: A Theology of Evolution.* Boulder: Westview.

————. 2004. Darwin, design, and the promise of nature. www.stmarylebow.co.uk/boyle%20lecture.htm.

Huxley, J. S. 1912. *The Individual in the Animal Kingdom.* Cambridge: Cambridge University Press.

————. 1959. Introduction to Teilhard de Chardin's *The Phenomenon of Man,* 11–28. London: Collins.

Medawar, P. [1961] 1967. Review of *The Phenomenon of Man.* In *The Art of the Soluble,* ed. P. Medawar. London: Methuen.

Rolston III, H. 1987. *Science and Religion.* New York: Random House.

————. 1999. *Genes, Genesis and God: Values and Their Origins in Natural and Human History.* Cambridge: Cambridge University Press.

————. 2001. Kenosis and nature. In *The Work of Love: Creation as Kenosis,* ed. J. Polkinghorne, 43–65. London: SPCK.

Ruse, M. 1993. Evolution and progress. *Trends in Ecology and Evolution* 8: 55–59.

Russell, R. J., W. R. Stoeger, and G. V. Coyne, eds. 1988. *Physics, Philosophy, and Theology: A Common Quest for Understanding.* Vatican City: Vatican Observatory.

Teilhard de Chardin, P. 1955. *The Phenomenon of Man.* London: Collins.

Thompson, D. W. 1917. *On Growth and Form.* Cambridge: Cambridge University Press.

Wagar, W. 1972. *Good Tidings: The Belief in Progress from Darwin to Marcuse.* Bloomington: Indiana University Press.

Ward, K. 1996. *God, Chance and Necessity.* Oxford: Oneworld.

Weber, B., and D. Depew. 2004. Darwinism, design, and complex system dynamics. In *Debating Design: Darwin to DNA,* ed. W. Dembski and M. Ruse, 173–190. Cambridge: Cambridge University Press.

Webster, J., ed. 2000. *The Cambridge Companion to Karl Barth.* Cambridge: Cambridge University Press.

12. Earth's Last Days?

Behe, M. 1996. *Darwin's Black Box: The Biochemical Challenge to Evolution.* New York: Free Press.

Boyer, P. 1992. *When Time Shall Be No More: Prophecy Belief in Modern American Culture.* Cambridge: Belknap Press.

Brooks, D. J. 2001. Substantial numbers of Americans continue to doubt evolution as explanation for origin of humans. Poll, The Gallup Organization.

DeHaan, M. R. 1962. *Coming Events in Prophecy.* Grand Rapids: Zondervan.

Dembski, W. A. 1998a. *The Design Inference: Eliminating Chance through Small Probabilities.* Cambridge: Cambridge University Press.

————, ed. 1998b. *Mere Creation: Science, Faith and Intelligent Design.* Downers Grove, IL: Intervarsity Press.

————. 1999. *Intelligent Design: The Bridge between Science and Theology.* Downers Grove, IL: Intervarsity Press.

————, 2001. Signs of intelligence: a primer on the discernment of intelligent design. *Signs of Intelligence: Understanding Intelligent Design,* ed. W. A. Dembski, and J. M. Kusiner. Grand Rapids: Brazos.

————. 2002. *No Free Lunch: Why Specified Complexity Cannot be Purchased without Intelligence.* Lanham, MD: Rowman & Littlefield.

Forrest, B., and P. R. Gross. 2004. *Creationism's Trojan Horse: The Wedge of Intelligent Design.* Oxford: Oxford University Press.

Frykholm, A. J. 2004. *Rapture Culture: "Left Behind" in Evangelical America.* Oxford: Oxford University Press.

Gish, D. 1973. *Evolution: The Fossils Say No!* San Diego: Creation-Life.

Graham, B. 1965. *World Aflame.* New York: Doubleday.

Holden, C. 2004. Georgia backs off a bit, but in other states battles heat up. *Science* 303: 1268.

Johnson, P. E. 1991. *Darwin on Trial.* Washington, DC: Regnery Gateway.

————. 1995. *Reason in the Balance: The Case against Naturalism in Science, Law and Education.* Downers Grove, IL: Intervarsity Press.

————. 1997. *Defeating Darwinism by Opening Minds.* Downers Grove, IL: Intervarsity Press.

————. 2000. *The Wedge of Truth: Splitting the Foundations of Naturalism.* Downers Grove, IL: Intervarsity Press.

Lindsey, H. 1970. *The Late Great Planet Earth.* Grand Rapids: Zondervan.

Morris, H. M. 1983. *The Revelation Record: A Scientific and Devotional Commentary on the Prophetic Book of the End of Times.* Wheaton, IL: Tyndale House.

————. 1989. *The Long War against God: The History and Impact of the Creation/Evolution Conflict.* Grand Rapids: Baker Book House.

Morris, H. M., et al. 1974. *Scientific Creationism.* San Diego: Creation-Life Publishers.

Numbers, R. L. 1992. *The Creationists: The Evolution of Scientific Creationism.* New York: Knopf.

————. 2004. Ironic heresy: how young-Earth creationists came to embrace rapid microevolution by means of natural selection. In *Darwinian Heresies,* ed. A. Lustig, R. Richards, and M. Ruse, 84–100. Cambridge, Cambridge University Press.

Overton, W. R. [1982] 1988. United States District Court Opinion: McLean versus Arkansas. In *But Is It Science? The Philosophical Question in the Creation/Evolution Controversy,* ed. M. Ruse 307–331. Buffalo: Prometheus.

Plantinga, A. 1991. An evolutionary argument against naturalism. *Logos* 12: 27–49.

Price, G. M. 1902. *Outlines of Modern Christianity and Modern Science.* Oakland: Pacific Press.

————. 1923. *The New Geology.* Mountain View: Pacific Press.

————. n.d. *Why I Am Not an Evolutionist.* Mountain View: Pacific Press.

Ruse, M., ed. 1988. *But Is It Science? The Philosophical Question in the Creation/Evolution Controversy.* Buffalo: Prometheus.

Shanks, N. 2004. *God, the Devil, and Darwin.* Oxford: Oxford University Press.

Smocovitis, V. B. 1999. The 1959 Darwin centennial celebration in America. *Osiris* 14: 274–323.

Tax, S., ed. 1960a. *Evolution after Darwin: The Evolution of Life.* Chicago: University of Chicago Press.

————, ed. 1960b. *Evolution after Darwin: The Evolution of Man.* Chicago: University of Chicago Press.

Tax, S., and Charles Callender, eds. 1960. *Evolution after Darwin: Issues in Evolution.* Chicago: Chicago University Press.

Webb, G. E. 1994. *The Evolution Controversy in America.* Lexington: University Press of Kentucky.

Whitcomb, J. C., and H. M. Morris. 1961. *The Genesis Flood: The Biblical Record and Its Scientific Implications.* Philadelphia: Presbyterian and Reformed Publishing Company.

Conclusion

Budd, S. 1977. *Varieties of Unbelief: Atheists and Agnostics in English Society 1850–1860.* London: Heinemann.

Camp, A. 1999. The intelligent design movement: an ally? *Creation Matters* 4 (6).

Darwin, C. 1959. *The Origin of Species by Charles Darwin: A Variorum Text*, ed. M. Peckham. Philadelphia: University of Pennsylvania Press.

Dawkins, R. 2003. *A Devil's Chaplain: Reflections on Hope, Lies, Science and Love.* Boston: Houghton Mifflin.

Dembski, W. A. 1999. *Intelligent Design: The Bridge between Science and Theology.* Downers Grove, IL: Intervarsity Press.

Dennett, D. C. 1995. *Darwin's Dangerous Idea.* New York: Simon and Schuster.

Forrest, B., and P. R. Gross. 2004. *Creationism's Trojan Horse: The Wedge of Intelligent Design.* Oxford: Oxford University Press.

Gish, D. T., A. Tsugita, J. Young, H. Fraenkel-Conrat, C. A. Knight, and W. M. Stanley. 1960. The complete amino acid sequence of the protein of tobacco mosaic virus. *Proceedings of the National Academy of Sciences* 46: 1463.

Gould, S. J. 1999. *Rocks of Ages: Science and Religion in the Fullness of Life.* New York: Ballantine.

Hamilton, W. D. 2001. *Narrow Roads of Gene Land: The Collected Papers of W. D. Hamilton.* Vol. 2, *Evolution of Sex.* Oxford: Oxford University Press.

Johnson, P. E. 1995. *Reason in the Balance: The Case against Naturalism in Science, Law and Education.* Downers Grove, IL: Intervarsity Press.

Livingstone, D. N. 1996. Evolution and eschatology. *Themelios* 22: 26–36.

Morris, J. D. 2002. Cracks are widening in evolution's dam! *Acts and Facts* 31 (5).

Plantinga, A. 1997. Methodological naturalism. *Perspectives on Science and Christian Faith* 49 (3): 143–154.

Popper, K. R. 1972. *Objective Knowledge.* Oxford: Oxford University Press.

———. 1974. Intellectual autobiography. In *The Philosophy of Karl Popper*, ed. P. A. Schilpp, vol. 1: 3–181. LaSalle, IL: Open Court.

Ruse, M. 1999. *Mystery of Mysteries: Is Evolution a Social Construction?* Cambridge: Harvard University Press.

———. 2001. *Can a Darwinian Be a Christian? The Relationship between Science and Religion.* Cambridge: Cambridge University Press.

Wilson, E. O. 2002. *The Future of Life.* New York: Vintage Books.

Acknowledgments

For twenty-five years now I have been looking at the relationship between science and the values of the cultures within which it is produced. I have focused on evolutionary thinking, and this has led to the writing of three books, all published by Harvard University Press: *Monad to Man: The Concept of Progress in Evolutionary Biology* (1996); *Mystery of Mysteries: Is Evolution a Social Construction?* (1999); and *Darwin and Design: Does Evolution Have a Purpose?* (2003). This present book, *The Evolution–Creation Struggle*, is rather less than a popular overview of my trilogy on science and values, and at the same time rather more. Less because, although I have used the findings of the earlier books, I have made no attempt to cover comprehensively what I found and argued in those books. More, because I wanted to use the conclusions that I drew in the trilogy to examine an issue that has engaged me even longer than the study of science and its values: the persistence and bitterness of the clash, especially in America, between evolutionists and those Christians who insist on non-natural origins of organisms. For me, as one born in England and raised in a Christian group (the Quakers) that welcomed the findings of science, the clash has always seemed very strange. While I never hesitated about the side to which I belong, as a philosopher of science I wanted to find out why there is this struggle. It is only now, after nearly a lifetime's thinking about evolution and its history, that I believe I am close to the answer.

As always when writing a book, I am very much in the debt of others. I could not have completed this project without the labors of two great Christian historians of American religion, George Marsden and Mark Noll. Their writings make me feel very humble and grateful. At a more personal level, I have gained much from the friendship, insights, writings, and criticisms of Ronald Numbers and David Livingstone. They showed me that my story would be radically incomplete without sensitivity to the significance of millennial thinking. My colleague John Corrigan also provided much useful information, and Miller Williams kindly allowed me to use his wonderful kangaroo poem. As will be seen from my text, I am particularly obliged to many of today's combatants, on both sides of the divide, for patiently answering my questions and allowing me to quote their replies. Never have our religious and other differences stood in the way of a courteous and friendly exchange of ideas.

For this book, I am even more in debt than usual to my sponsoring editor at Harvard University Press, Michael Fisher. He read several drafts and helped me to shape the content and sharpen the arguments that lead to my conclusion. The Press's anonymous referees also gave terrific suggestions for improvement, for which I am grateful. And let the reader be glad that nothing I wrote saw the light of day until it was repaired and renovated by my incomparable manuscript editor, Susan Wallace Boehmer.

Finally, I want to thank my wife, Lizzie, for her love and support, and to say with what pleasure I dedicate this book to my long-time friend, the philosopher and historian of science Bob Richards. He rarely agrees with anything that I have written, and I shall be disappointed if he starts now.

Index